Revise GCSE

Physics

Contents

This book and your GCSE course

<table>
<tr><th></th><th>AQA</th><th>EDEXCEL 360</th></tr>
<tr><td>Web address</td><td>www.aqa.org.uk</td><td>www.edexcel.org.uk</td></tr>
<tr><td>Specification number</td><td>4451</td><td>2109</td></tr>
<tr><td>Modular tests</td><td>either
P1 Structured questions, 45 mins, 25%
or
P1a and P1b
Matching/multiple choice Questions, 30 mins each, 25%
P2 short answer questions, 45 mins, 25%
P3 short answer questions, 45 mins, 25%</td><td>P1a 10% and P1b Multiple choice
20 mins each, 10%
P2 structured test 10%,
multiple choice 10%
P3 either structured questions
or portfolio 30%</td></tr>
<tr><td>Terminal papers</td><td>none</td><td>none</td></tr>
<tr><td>Availability of exams</td><td><table><tr><th></th><th>P1</th><th>P1a P1b</th><th>P2 P3</th></tr><tr><td>Nov</td><td></td><td>✓</td><td></td></tr><tr><td>Jan</td><td>✓</td><td></td><td>✓</td></tr><tr><td>March</td><td></td><td>✓</td><td></td></tr><tr><td>June</td><td>✓</td><td>✓</td><td>✓</td></tr></table></td><td>All exams available in November
March and June except P3
only in June</td></tr>
<tr><td>Coursework</td><td>25%</td><td>30%</td></tr>
<tr><td>PHYSICS</td><td></td><td></td></tr>
<tr><td>Radioactivity</td><td>P1.11.4, P1.11.5, P1.11.6, P2.12.9, P2.12.10</td><td>P1b11, P2.11, P2.12, P3.5</td></tr>
<tr><td>Energy</td><td>P1.11.1, P1.11.2, P1.11.3, P1.11.4, P1.11.5, P2.12.7, P2.12.8, P3.13.8, P3.13.9</td><td>P1a9, P1a10, P2.12</td></tr>
<tr><td>Waves</td><td>P1.11.5, P3.13.4, P3.13.5, P3.13.6</td><td>P1b11, P3.6</td></tr>
<tr><td>Electromagnetic radiation</td><td>P1.11.5, P1.11.6</td><td>P1b11, P3.6</td></tr>
<tr><td>The Earth and beyond</td><td>P1.11.7, P3.13.3, P3.13.10</td><td>P1b12, P2.11</td></tr>
<tr><td>Forces and motion</td><td>P2.12.1, P2.12.2, P2.12.3, P2.12.4</td><td>P1b12, P2.9, P2.10, P3.6</td></tr>
<tr><td>Electricity</td><td>P2.12.5, P2.12.6, P2.12.7, P2.12.8, P3.13.8, P3.13.9</td><td>P1a9, P1a10, P2.12, P3.6</td></tr>
<tr><td>More waves and radiation</td><td>P3.13.4, P3.13.5, P3.13.6</td><td>P2.11, P3.6</td></tr>
<tr><td>Nuclear physics</td><td>P1.11.6, P2.12.9, P2.12.10</td><td>P2.9, P2.11, P2.12, P3.5, P3.6</td></tr>
<tr><td>Particles</td><td></td><td>P3.5</td></tr>
<tr><td>More forces and motion</td><td>P2.12.4, P3.13.1, P3.13.2, P3.13.3</td><td>P2.9, P3.6</td></tr>
<tr><td>Stars and the Universe</td><td>P3.13.4, P3.13.10</td><td>P3.5</td></tr>
<tr><td>Applications – electronics and medicine</td><td>P3.13.7, P3.13.8</td><td>P1a10, P3.6</td></tr>
</table>

Visit your awarding body for full details of your course or download your complete GCSE specifications

Use these pages to get to know your course

- ***Make sure you know your exam board***
- ***Check which specification you are doing***
- ***Know how your course is assessed:***
  - ***– what format are the papers?***
  - ***– how is coursework assessed?***
  - ***– how many papers?***

OCR A	OCR B
www.ocr.org.uk	
J635	J645
Unit 1 Modules P1, P2 and P3 Objective style questions, 40 mins, 16.7% Unit 2 Modules P4, P5 and P6 Objective style questions, 40 mins, 16.7% Unit 3 Ideas in Context + P7 Structured questions, 40 mins, 33.3% *either* Unit 4 Practical Data Analysis and Case Study *or* Unit 5 Practical Investigation 33.3%	Unit 1 Modules P1, P2 and P3 Structured questions, 60 mins, 33.3% Unit 2 Modules P4, P5 and P6 Structured questions, 60 mins, 33.3% *either* Unit 3 'Can-do tasks and report on 'Science in the News' *or* Unit 4 Research Study, Data Tasks and Practical Skills 33.3% Terminal Papers none none Unit 3 none Availabilty of exams
Unit 3	none

OCR A:

	U1	U2	U3	U4	U5
Jan	✓	✓			
June	✓	✓	✓	✓	✓

OCR B:

	U1	U2	U3	U4
Jan	✓	✓		
June	✓	✓	✓	✓

OCR A	OCR B
33.3%	33.3%
P1.1, P2.1, P2.2, P2.5, P3.1, P3.2, P3.3, P3.4, P7.4	P2c, P2d, P4e, P4f, P4g, P4h
P2.5, P3.3, P3.4, P5.4, P5.5	P1a, P1b, P1c, P2a, P2b, P2c, P6d, P6e
P1.4, P6.1, P6.2, P6.3	P1d, P1e, P1f, P1g, P1h, P5f, P5g
P2.1, P2.2, P2.3, P2.4, P2.5, P3.2, P6.3, P6.4	P1d, P1e, P1f, P1g, P1h, P4d, P4e, P5e
P1.1, P1.2, P1.3, P1.4, P7.3, P7.4	P2e, P2f, P2g, P2h, P5a
P4.1, P4.2, P4.3, P4.4	P3a, P3b, P3c, P3d, P3e, P3f, P3g, P3h, P5a, P5d
P5.1, P5.2, P5.3, P5.4, P5.5	P4a, P4b, P4c, P6a, P6b, P6d, P6e, P6f
P6.1, P6.2, P6.3, P6.4	P4d, P4e, P5e, P5f, P5g
P3.4, P7.4	P4e, P4f, P4g
P7.4, P7.5	
P4.3	P5a, P5b, P5c, P5d
P7.1, P7.2, P7.3, P7.4, P7.5	P5h
P3.4	P6a, P6b, P6c, P6d, P6f, P6g, P6h

Preparing for the examination

Planning your study

The last few months before taking your final GCSE examinations are very important in achieving your best grade. However, the success can be assisted by an organised approach throughout the course. This is particularly important now as all the science courses are available in units.

- After completing a topic in school or college, go through the topic again in your Revise GCSE Physics Study Guide. Copy out the main points on a sheet of paper or use a highlighter pen to emphasise them.
- Much of memory is visual. Make sure your notes are laid out attractively using spaces and symbols. If they are easy to read and attractive to the eye, they will be easier to remember.
- A couple of days later, try to write out these key points from memory. Check differences between what you wrote originally and what you wrote later.
- If you have written your notes on a piece of paper, keep this for revision later.
- Try some questions in the book and check your answers.
- Decide whether you have fully mastered the topic and write down any weaknesses you think you have.

Preparing a revision programme

Before an external examination, look at the list of topics in your examination board's specification. Go through and identify which topics you feel you need to concentrate on. It is a temptation at this time to spend valuable revision time on the things you already know and can do. It makes you feel good but does not move you forward.

When you feel you have mastered all the topics, spend time trying sample questions that can be found on your examination board's website. Each time, check your answers with the answers given. In the final week, go back to your summary sheets (or highlighting in the book).

How this book will help you

Revise GCSE Physics Study Guide will help you because:

- it contains the essential content for your GCSE course without the extra material that will not be examined
- it contains GCSE Exam Practice Questions to help you to confirm your understanding
- examination questions from 2007 are different from those in the past. Trying past questions will not help you when answering some parts of the questions in 2007. The questions in this book have been written by experienced examiners.
- the summary table will give you a quick reference to the requirements for your examination

Five ways to improve your grade

1. Read the question carefully

Many students fail to answer the actual question set. Perhaps they misread the question or answer a similar question they have seen before. Read the question once right through and then again more slowly. Some students underline or highlight key words in the question as they read it through. Questions at GCSE contain a lot of information. You should be concerned if you are not using the information in your answer.

2. Give enough detail

If a part of a question is worth three marks you should make at least three separate points. Be careful that you do not make the same point three times. Draw diagrams with a ruler and label with straight lines. For questions where you choose true or false statements, or tick boxes, there are marks for correctly identifying true *and* false statements so, for example, if there are 3 marks this does not mean that there are three true answers.

3. Correct use of scientific language

There is important scientific vocabulary you should use. Try to use the correct scientific terms in your answers. The way scientific language is used is often a difference between successful and unsuccessful students. As you revise, make a list of scientific terms you meet and check that you understand the meaning of these words. Learn all the definitions. These are easy marks and they reward effort and good preparation.

4. Show your working

All science papers include calculations. Learn a set method for solving a calculation and use that method. You should always show your working in full. Then, if you make an arithmetical mistake, you may still receive marks for correct science. Check that your answer is given to the correct number of significant figures and give the correct units.

5. Make sure your work is marked

Exam papers are often scanned and marked on-screen, so it is important to make sure you write only in the spaces allowed (or ask for extra paper.) Write 'See…' by any asterisks you use so the examiner knows there is extra work off-screen. Use a black pen and write clearly.

How Science Works

From 2007, all GCSE science courses must cover certain factual detail, similar to the detail that has been required for many years. Now, however, each course must also include study of '**How Science Works**'.

This includes four main areas:

- **Data, evidence, theories and explanations**
  This involves learning about how scientists work, the differences between data and theories and how scientists form theories.

- **Practical skills**
  How to test a scientific idea including collecting the data and deciding how reliable and valid it is.

- **Communication skills**
  Learn how to present information in graphs and tables and to be able to analyse information that has been provided in different forms.

- **Applications and implications of science**
  Learning about how new scientific discoveries become accepted and some of the benefits, drawbacks and risks of new developments.

The different examining bodies have included material about how science works in different parts of their examinations. Often it is in the coursework but you are also likely to come across some questions in your written papers. Do not panic about this and think that you have not learnt this work. Remember these questions test your skills and not your memory; that is why the situations are likely to be unfamiliar. The examiners want you to show what you know, understand and can do.

To help you with this, there are sections at the end of each chapter called **How Science Works** and questions about how science works in the **Exam Practice Questions**. This should give you an idea of what to expect.

These ideas are examined throughout the course, so although the '**How Science Works**' sections are at the end of each chapter, all of them cover ideas that you may find useful in the early module tests, as well as later tests. Read through them early in your course.

Chapter 1

Radioactivity

The following topics are covered in this chapter:

- *Radioactive emissions*
- *Dangers of radiation*
- *Uses of radioactivy*
- *Changes in the nucleus*
- *Nuclear reactors*

1.1 Radioactive emissions

Alpha, beta and gamma radiation

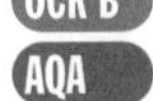

OCR A P3.1
OCR B P2d
AQA P1.11.6
EDEXCEL P2.11,3.5

A **radioactive nucleus** is **unstable** and will emit radiation. There are three main types:

- **Alpha** (α) is strongly ionising radiation, but it only travels a few centimetres in air and is stopped by a thin sheet of paper.
- **Beta** (β) is ionising radiation that penetrates card or several sheets of paper, but is stopped by a 3mm thick sheet of aluminium or other metal.
- **Gamma** (γ) is weakly ionising radiation that is very penetrating. It is reduced significantly by a thick lead sheet or blocks of concrete.

KEY POINT

There are three types of radioactive emissions: alpha, beta and gamma radiation. Alpha radiation is the most ionising, gamma radiation is the most penetrating.

Alpha and **beta** radiation is deflected by **magnetic fields** and **electric fields**, but **gamma** radiation is not affected.

Background radiation

OCR A P3.2
OCR B P2d,4g
AQA P2.12.9
EDEXCEL 360 P2.11

Background radiation is radioactive emissions from nuclei in our surroundings. Do not confuse this with radiation from other sources, such as radiation from mobile phones.

Radioactive materials occur naturally, and can also be made artificially. **Cosmic rays** from space make some of the carbon dioxide in the atmosphere radioactive. The carbon dioxide is used by plants and enters food chains. This makes all living things radioactive. Some rocks are also radioactive.

We receive a low level of radiation from these sources all the time. It is called **background radiation.** Background radiation comes from:

- space (cosmic rays) and the Sun
- building materials, rocks (e.g. granite) and soil
- radioactive nuclei in all plants and animals
- medical and industrial uses of radioactive materials
- 'leaks' from radioactive waste and nuclear power stations.

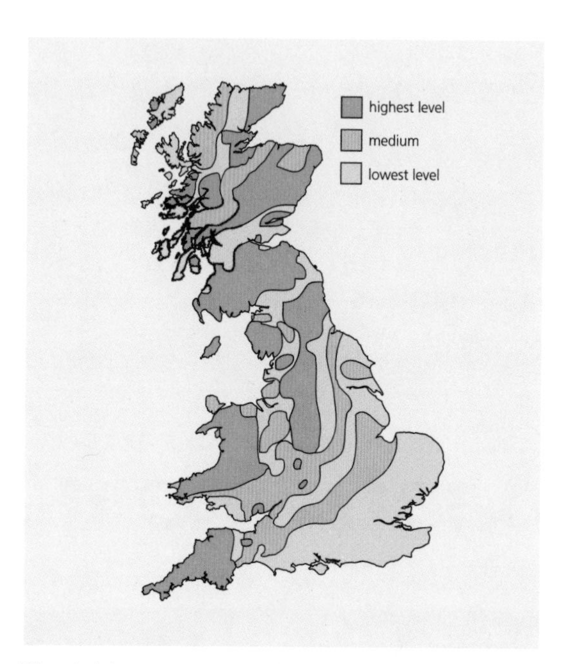

Fig. 1.1 In some parts of the country the rocks are more radioactive than in others and there is a higher level of background radiation.

1.2 Dangers of radiation

Ionising radiation

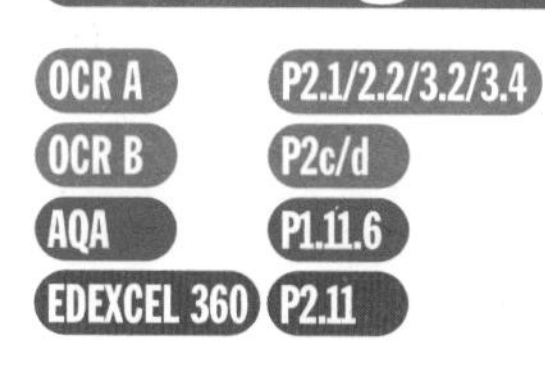

Ionising radiation is radiation that has enough energy to break molecules or atoms into charged particles called **ions**. The molecules or atoms lose **electrons** in a process called **ionisation**. The ions can then take part in chemical reactions. This is how ionising radiation damages **living cells**. It kills them, or damages the **DNA** in the cell so that the cell **mutates** (changes) into a **cancer** cell, which then grows in an uncontrolled manner.

Gamma rays, X-rays and ultra-violet radiation are the three types of electromagnetic radiation with high enough frequency to cause ionisation.

Ionising radiation includes alpha, beta and gamma radiation from the radioactive nuclei and also X-rays and ultra-violet radiation.

Of the radioactive emissions, alpha particles are the most ionising, followed by beta, and gamma are the least ionising.

Contamination or irradiation?

If a person has a radioactive material on their skin or clothes, or has swallowed or inhaled it, this is called **radioactive contamination**. The radioactive material will decay over a period of time. This results in a higher dose of radiation than being exposed to an external source, which is called **irradiation**, because when the person moves away from the source they are no longer irradiated. This is especially true of the most ionising radiation, alpha radiation. Unlike beta and gamma radiation, alpha radiation cannot penetrate the skin, so outside the body – a few centimetres away – it is not dangerous. But if a person inhales, or swallows, material that emits alpha particles, the source will continue to emit them inside the body until all the radioactive nuclei have decayed. This increases the risk of cancer.

Make sure you know the difference.

Risk and safety

OCR A P2.5/3.1/3.2/3.4
OCR B P2d
AQA P1.11.6
EDEXCEL 360 P2.11

It is **not possible to predict** which cells will be damaged by exposure to radiation, and it is not possible to say who will get cancer. Scientists have studied the survivors of incidents where people have been exposed to ionising radiation. They measured the amount of exposure and recorded how many people later suffered from cancer.

They introduced **radiation dose**, measured in **sieverts**, which is a measure of the **possible harm done to the body**. Radiation dose depends on the **type of radiation**, the **time of exposure** and how **sensitive the tissue** exposed is to radiation. Scientists can then give a figure for the **risk** of cancer developing.

To reduce the risk to living cells, radioactive materials must be handled **safely**:

- wear protective clothing
- keep a long distance away (use tongs to handle sources)
- keep the exposure time short
- sources should be shielded and labelled with the radioactive symbol.

Fig. 1.2 Radioactive hazard symbol.

These precautions keep the dose as low as possible. Radioactive materials are used, for example, in hospitals and nuclear power stations. Employers must keep the dose for their employees **a**s **l**ow **a**s **r**easonably **a**chievable: this is known as the **alara** principle.

You may have had an X-ray photograph taken and noticed that the staff move behind a screen or out of the room, so that they are not exposed to X-rays every time the X-ray machine is used. Employees at nuclear powers stations, and other places where nuclear sources are used, may wear a **film badge** to monitor the exposure to radiation. Airline employees working as flight crew are monitored too, because the exposure to cosmic rays is increased if you fly at high altitudes in the thin atmosphere of the Earth.

1.3 Uses of radioactivity

Uses of radioactive sources

OCR A P3.2
OCR B P2d,4e
AQA P1.11.6
EDEXCEL 360 P2.11

For each use:

- the radiation is chosen depending on the range and the absorption (see 1.1 Alpha, beta and gamma radiation)
- the source is chosen depending on how long it will remain radioactive (see 1.4 Radioactive decay and Half-life)

Medical and health uses

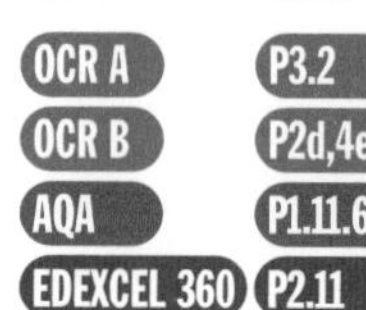

Medical tracers

Sources which emit **gamma radiation** (or sometimes **beta radiation**) are used in **medical tracers**. The patient drinks, inhales, or is injected with the tracer which is chosen to target the organ doctors want to examine. For example, radioactive iodine is taken up by the thyroid gland, which can then be viewed using a **gamma camera** that detects the gamma radiation passing out of the body. The tracer must not decay before it has moved to the organ being investigated but it must not last so long that the patient stays radioactive for weeks afterwards. Sources which emit a higher dose of **gamma radiation** are used in the same way for **treating cancer** by building up in the cancer and killing the cancer cells.

Treating cancer

A beam of gamma rays is directed at a cancer to kill the cancer cells. One source used for this is cobalt-60. It emits high energy gamma rays and remains radioactive for years.

Sterilisation

Sources which emit **gamma radiation** are used to produce a beam of gamma rays that will:

- **sterilise equipment**, such as surgical instruments, by destroying microbes
- extend the shelf-life of perishable **food**, by destroying microbes.

The food and equipment does not become radioactive because it is only **irradiated**. It does not touch the radioactive material, so there is no **contamination.**

Other uses of radioactive sources

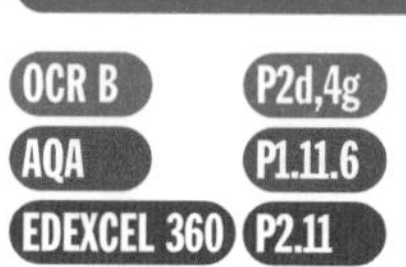

Smoke detectors

Radioactive sources that emit **alpha radiation** are used in **smoke detectors**. The alpha radiation from the source ionises the air and the ions cross a small gap and are picked up by a detector. If smoke is present, the alpha radiation is stopped by the smoke particles. No ions reach the detector and an alarm is sounded.

Beta and gamma radiation are unsuitable because they pass through the smoke.

Tracers

Sources which emit **beta radiation** or **gamma radiation** are used as **tracers**. Because a tracer is radioactive, detectors can be used to track where it goes. If a tracer is added to sewage at an ocean outlet, or as it enters a river, then the movement can be traced. Leaks in power station heat exchangers can be tracked. The source used is carefully selected to be one with a radioactivity that will fall to zero quickly after the test is done.

Paper thickness detectors

Radioactive sources that emit **beta radiation** are used in **paper thickness detectors**. Fig. 1.3 shows how the paper thickness can be monitored. Some of the beta radiation is **absorbed** by the paper sheet. If the sheet is too thick, less beta radiation is detected and the pressure of the rollers is increased. If the sheet is too thin more beta radiation is detected and the pressure is reduced.

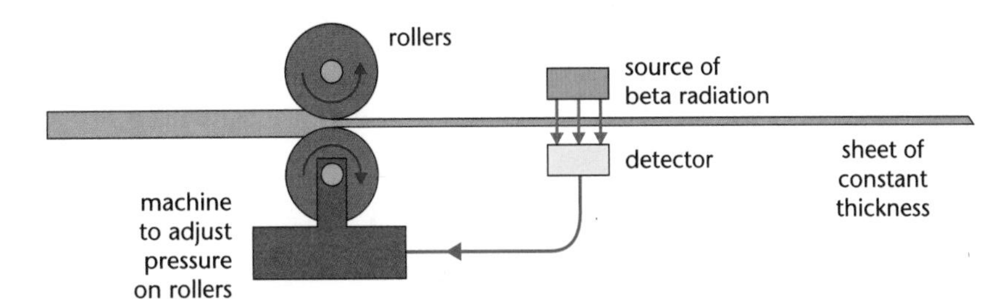

Fig. 1.3 Using a beta source to control paper thickness.

Non-destructive testing

Another use of **gamma radiation** sources is **non-destructive testing**. An aircraft wing can be examined for minute cracks by placing a strong gamma source on one side and a detector on the other – in a similar way to using X-rays to check for a broken bone.

1.4 Changes in the nucleus

The nucleus and isotopes

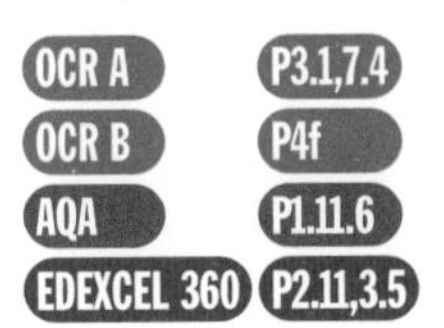

When an atom loses electrons it becomes a positively charged ion. If it gains electrons it becomes a negatively charged ion.

The **atom** is mostly empty space with almost all the **mass** concentrated in the small **positively charged nucleus** at the centre. The **nucleus** of an atom contains two types of particle:

- **neutrons**, which have no charge
- **protons** – each proton has a single positive charge.

The **nucleus** is surrounded by orbiting **electrons** which have very little mass. Each **electron** has a negative charge. A neutral atom has an equal number of protons and electrons.

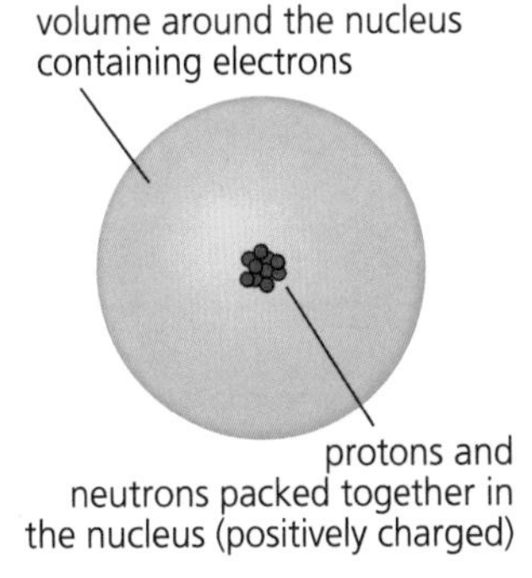

Fig. 1.4 Arrangement of neutrons, protons and electrons in an atom.

Different elements

The number of **protons** in the nucleus decides which **element** the atom is, so for example, hydrogen always has one proton, helium has two, lithium has three and so on. Carbon always has six.

Isotopes

The number of neutrons in the nucleus can vary. An **isotope** is a nucleus of an element with the **same number** of **protons** but different **number** of **neutrons**. Some isotopes are stable but others are unstable. Unstable isotopes are radioactive because they emit nuclear radiation in the process called **radioactive decay**.

Isotopes of an element have the same number of protons in the nucleus, but different numbers of neutrons.

Isotopes of the same element have exactly the same chemical properties, but they have different density and nuclear stability. Different isotopes are referred to by the number of nucleons in the nucleus, for example, carbon-12 has 6 protons and 6 neutrons, whereas carbon-14 has 6 protons and 8 neutrons.

Radioactive decay

A radioactive nucleus is unstable and emits nuclear radiation. This process is called **radioactive decay**. It is not possible to predict when this will happen, nor is it possible to make it happen by a chemical or physical process, (for example by heating it.) The **decay** is **random**.

Radioactive emissions

Alpha emission is when **two protons and two neutrons** leave the nucleus as one particle. The alpha particle is identical to a helium nucleus. For example, the isotope of radon gas, radon-220, decays by alpha emission.

radon-220 $\rightarrow$ polonium-216 + alpha particle

Beta emission occurs when a **neutron** decays to a **proton** and an **electron** and the electron leaves the nucleus. The beta particle is a high energy electron.

The radioactive isotope of carbon, carbon-14 decays by beta emission.

carbon-14 $\rightarrow$ nitrogen 14 + beta particle

A beta particle is an electron from the nucleus – not an orbital electron.

An alpha particle is two protons and two neutrons – a helium nucleus. A beta particle is a high-energy electron from the nucleus. Gamma radiation is a high-frequency and short wavelength electromagnetic wave.

Half-life

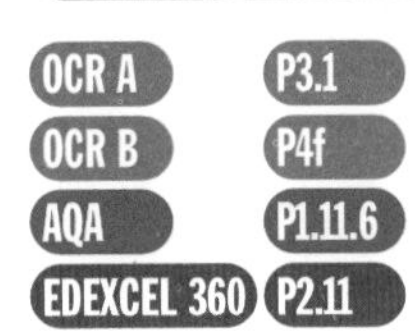

A radioactive source contains millions of nuclei. The number of nuclei decaying per unit time is called the **activity** of the source. The activity depends on two things:

- the type of isotope – some isotopes are more stable than others.
- the number of undecayed nuclei in the sample – double the number of nuclei and, on average, there will be double the number of decays per second.

Over a period of time the activity of a source gradually dies away.

> **KEY POINT** **The half-life of an isotope is the average time taken for half of the active nuclei to decay.**

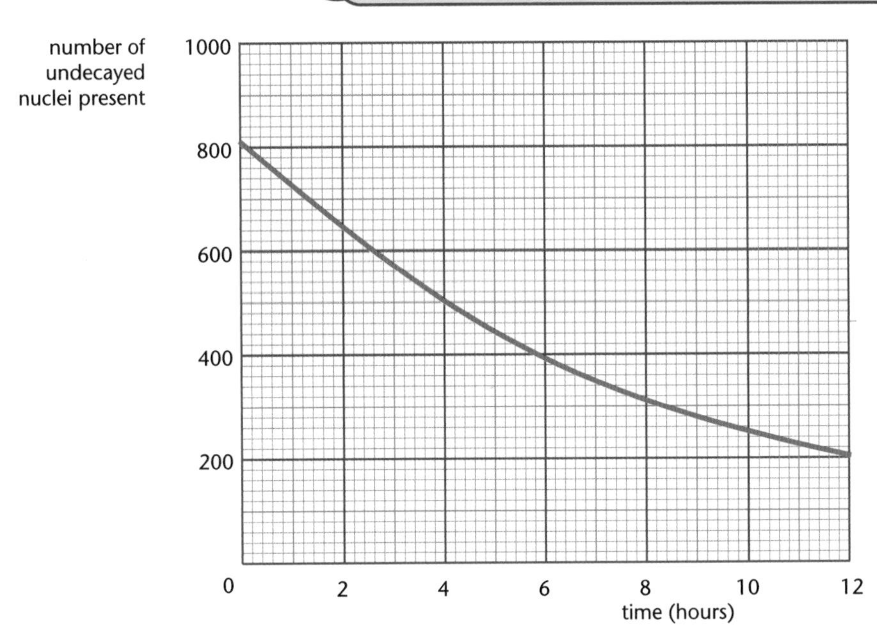

Fig. 1.5 The decay of a sample radioactive nuclei with a half-life of six hours.

Technetium-99m (Tc-99m) decays by gamma emission to technetium-99 (Tc-99) with a half-life of six hours. After six hours, on average, only half of the Tc-99m nuclei remain active. After another six hours, on average, only one quarter of the nuclei are active. This is shown on the graph in Fig. 1.5. This pattern is the same for all isotopes but the value of the half-life is different. Carbon-14 has a half-life of 5730 years, but some isotopes have a half-life of less than a second.

Tc-99m is one of the most widely used radioactive isotopes in medicine. It is used to diagnose problems in many organs. The half-life of six hours means that the radiation lasts only long enough for the isotope to travel to the organ being investigated. Its activity decreases rapidly and cannot be detected after a few days. Tc-99m does not occur naturally – it is made in reactors.

Dating

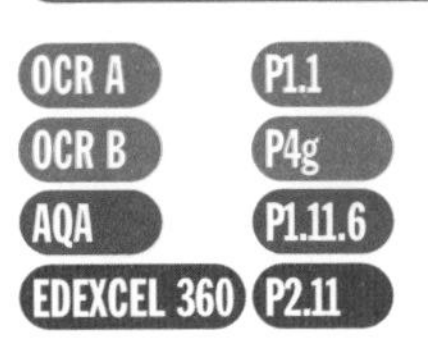

The amount of carbon-14 in a living object is fixed when it dies. The carbon-14 decays, so that a wooden spear that is 5730 years old has only half the carbon-14 left that it had when it was made. Objects that once lived can be carbon dated by the amount of carbon-14 that is left.

Some rocks contain a radioactive isotope of uranium that decays to lead, so they can be dated by the uranium – lead ratio, the more lead there is the older the rock is.

Because of the small amounts of isotopes involved, and the long half-lifes, these methods cannot be used to find dates to within tens of years. They are not useful for objects that are less than a hundred years old.

1.5 Nuclear reactors

Nuclear fission

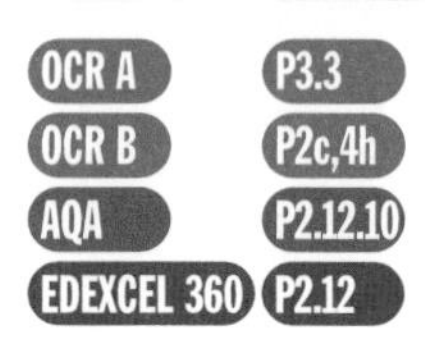

If a nucleus of uranium-235 absorbs a neutron, it becomes very unstable and can split into two nuclei of about equal size, and two or three neutrons. This process is called **nuclear fission**.

When this happens a lot of nuclear energy is released, about a **million times more** than the energy released in a chemical reaction.

The neutrons released can strike more uranium nuclei and cause more fission reactions – which in turn produce more neutrons, and so on. This is called a **chain reaction**.

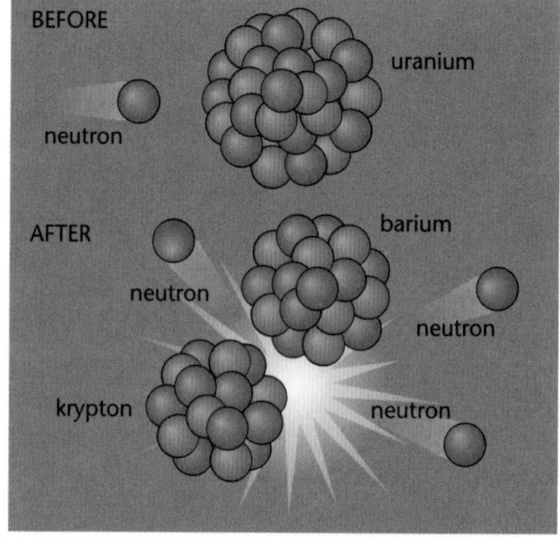

Fig. 1.6 A neutron absorbed by a uranium nucleus causes nuclear fission.

If the fission reaction runs out of control it is an atomic bomb, but if the process is controlled, the energy released can be used to generate electricity. This is how a **nuclear reactor** in a **power station** works.

The **fuel rods** contain uranium-235 and are put in the reactor. The reaction can be controlled or stopped by using a material that absorbs neutrons. The neutron absorbing material is made into **control rods**. The control rods are moved into the reactor to absorb neutrons, and slow or stop the reaction, and out of the reactor to increase the reaction.

The energy heats up the fuel rods and control rods. A **coolant** is circulated to remove the heat from the reactor. When the coolant has been heated it is used to heat water to steam for the power station. When the coolant is cool it circulates through the reactor again.

This diagram shows how one neutron striking a U-235 nucleus can set off a chain reaction.

Another isotope that can be used as a nuclear fuel is plutonium-239.

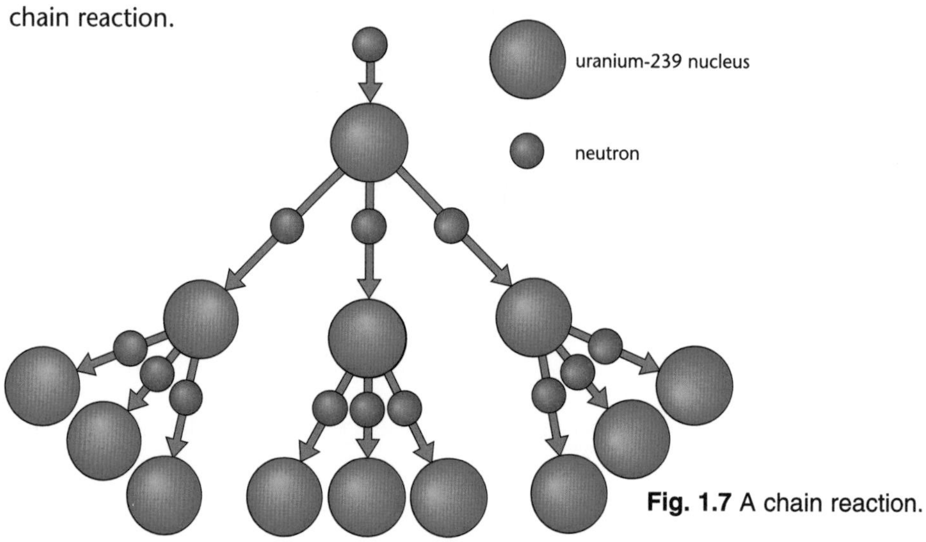

Fig. 1.7 A chain reaction.

Waste disposal

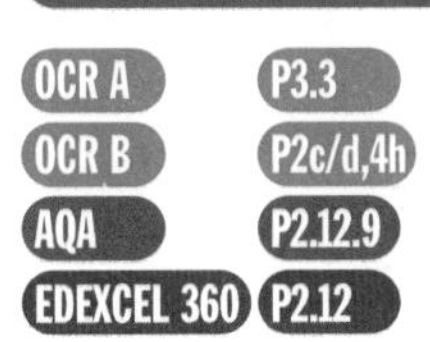

Radioactive waste is dangerous to living things and must be carefully disposed of. The half-life of some isotopes is thousands, or millions, of years, so radioactive material must be disposed of in a way that will keep it safely contained for thousands of years.

There are three types of radioactive waste:

- **low-level waste** – for example, used protective clothing.
- **intermediate-level waste** – for example, material from reactors.
- **high-level waste** – for example, used fuel rods.

Low-level waste can be sealed into containers and put in **landfill sites**.

Intermediate-level waste is mixed with concrete and **stored** in **stainless-steel containers**. It must be stored for thousands of years.

High-level waste is kept in **cooling tanks** at first because it decays so fast it gets hot. Eventually it becomes intermediate-level waste. High-level waste includes '**weapons grade plutonium**', which is the radioactive element plutonium, produced in nuclear reactors. It can be used to make nuclear weapons.

Waste can be **dispersed** (for example when sewage is discharged in the sea), or **contained** (for example when rubbish is put in a landfill site). Some radioactive waste is too dangerous to be dispersed.

Where to store the waste?

- At the bottom of the sea – but containers may leak.
- Underground – but containers may leak, and earthquakes or other changes to the rocks may occur.
- On the surface – but needs guarding (for example, from terrorists) for thousands of years.
- Blast into space – but danger of rocket explosion.

The **precautionary principle** says that if we are not sure of the effect of something, and it could be very harmful, then we should not risk trying it. (It is better to be safe than sorry.)

HOW SCIENCE WORKS

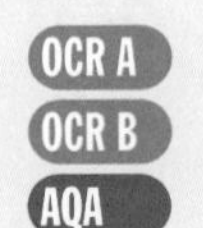

Investigating radon

In parts of the UK, the rocks contain small amounts of radioactive compounds, for example, uranium compounds. One of the decay products is radon gas, which is radioactive and decays by alpha emission. Radon gas that is breathed into the lungs can irradiate the lungs and cause cancer.

Do houses in some areas contain dangerous levels of radon gas?
Alpha particles leave tracks on a specially made plastic material, so a piece of the plastic can be placed in a house for a week and the number of tracks counted. This 'track count' can be used to compare the radioactivity due to alpha emission in different houses.

This experiment compared the track count for houses in a village where the rocks were known to emit radon.

The experiment was repeated three times in each house and the **mean** (**average**) was worked out. The data that is ringed in the table is obviously wrong. These **outliers**, (also called **anomalous results**) have been left out of the calculations. The **mean** is the best estimate of the value of the track count for each house. You can have **more confidence** in the mean than in a single measurement. The **range** of values, from the lowest to the highest, also gives you an idea of how **confident** you can be that the **mean** value is correct.

House	Age (years)	Track count (tracks per week)			
		1st test	2nd test	3rd test	mean
A	4	10	25	10	15
B	10	17	44	19	30
C	14	(125)	15	35	25
D	52	142	100	121	121
E	60	106	106	160	124
F	104	234	(105)	204	219
G	116	251	219	235	235
H	120	235	270	245	250

A graph was plotted of the mean track count against the age of the house.

You can be **confident** that there is a **real difference** in the count for house E and house F, because the **ranges do not overlap**. The difference between house F and house G could be just due to small fluctuations in the measurements – because the **ranges overlap** – you cannot be confident that there is a **real difference**. A set of data with large ranges is not as **reliable** as a set of data with small ranges.

Is age a **factor** in the number of alpha particles – the track count – in a house?

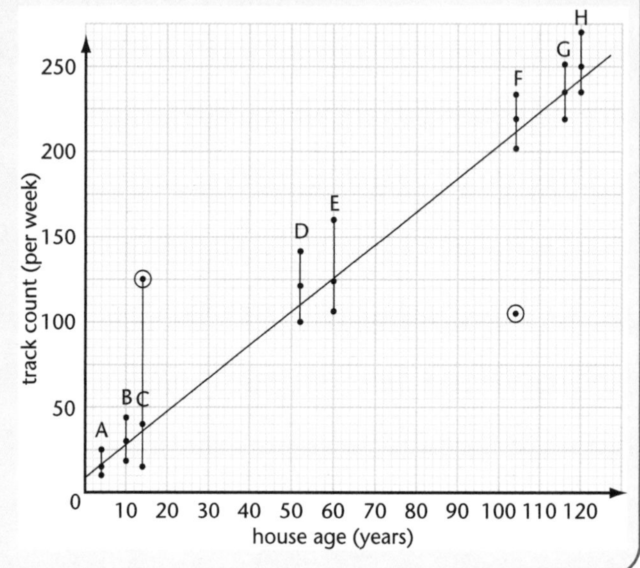

HOW SCIENCE WORKS

The greater the age, the greater the track count (the track count is the **outcome**). There is a **correlation** between the age of the house (the factor) and the track count (the outcome).

Does this mean that the house getting older is the **cause** of more alpha particle emission? The answer is no. When scientists compared the houses, they found that houses more than 70 years old were built of granite, with granite floors. The granite emits radon in addition to the rock on which the house is built. Houses less than 70 years old were built of brick, plastered, and with wooden floors, so radon still gets through from underground, but not from the walls and floors. Houses less than 15 years old have a gap and a fan under the floor, to move the gas outside. This removes the radon gas before it enters the house. This tells us that the **correlation** between track count and age of the house is not **caused** by age. It is another **factor** – the type of construction.

In another study, the colour of each front door was noted, and results showed that houses with yellow front doors had higher track counts. This was just **coincidence**. When more houses were investigated, yellow doors did not correlate with increased track count.

1. Choose from house A–H.
   (a) Which house had the largest range of track counts?
   (b) Which house had the smallest range of track counts? **[2]**
2. Explain whether there is a real difference between the track counts for house F and house H. **[1]**
3. Another house that was 70 years old had these track counts: 100, 150, 170 (tracks per week).
   Work out **(a)** the mean and **(b)** the range of the track counts. **[2]**
4. Write **True** or **False** for each statement.
   (a) There is a correlation between the method of house construction and the track count.
   (b) There is no correlation between the age of the house and the track count.
   (c) Increasing age was a factor that increased the track count.
   (d) The fact that increasing age correlated with increasing track count was coincidence.
   (e) Building construction was a factor that affected the track count
   (f) There is no real difference between the track counts for houses under 15 years old.
   (g) There is a real difference between the track counts for houses under 15 years old and those for houses between 50 and 60 years old. **[7]**

Exam practice questions

1. Nuclear radiation passes through a sheet of aluminium foil. The radiation could be:
   A alpha radiation only
   B beta radiation only
   C alpha, beta or gamma radiation
   D beta or gamma radiation **[1]**

2. Which of the following is a source of background radiation?
   A alpha radiation
   B microwave ovens
   C mobile phones
   D rocks **[1]**

3. A patient is given a dose of iodine-131 which has a half-life of eight days. What fraction of the iodine-131 nuclei are left after 32 days?
   A $\frac{1}{4}$
   B $\frac{1}{8}$
   C $\frac{1}{16}$
   D $\frac{1}{32}$ **[1]**

4. Use words from this list to complete the following sentence:

   electrons carbon hydrogen nuclei protons stable unstable

   Radioactive emissions come from the __________ of ________ atoms. **[2]**

5. Use the words below to complete this description of a nuclear reactor. (Use words once, more than once or not at all.)

   atom electron fission control rods coolant chain

   fuel rods neutron neutrons nucleus protons

   The uranium–235 is used to make the _______**1**_______. When a uranium _____**2**_____ is hit by a _____**3**_____ the nucleus splits into two smaller nuclei and a few more _____**4**_____. A lot of energy is released. This process is called nuclear _____**5**_____. The neutrons hit more uranium nuclei and produce a _____**6**_____ reaction. The reaction is slowed or stopped by moving _____**7**_____ into the reactor. These absorb _____**8**_____. The heat energy from the reactor is removed by the _____**9**_____ and used to heat water in the power station. **[9]**

Exam practice questions

6. Match the descriptions with the type of nuclear waste by drawing straight lines to link the boxes.

Description	Type of waste
used fuel from nuclear reactors	high level waste
reactor fuel cladding	intermediate level waste
the protective clothing worn by workers	low level waste

[2]

7. **(a)** The alara principle says that exposure to radioactive materials should be kept _______ _______ ______ _______ ________

(b) Bob says that high-level radioactive waste should be stored at the power station until a safe way to dispose of it is found. This is an example of the 'precautionary principle'. Explain what this means. **[3]**

8. **(a)** Which type of nuclear radiation is used in a paper thickness detector?

(b) Explain why this type is chosen. **[3]**

9. **(a)** A patient is injected with technetium-99m which has a half-life of 6 hours. What fraction of the technetium-99m nuclei are left after one day?

(b) Why is it better to use an isotope with a half-life of 6 hours rather than:

(i) an isotope with a half-life of 6 minutes

(ii) an isotope with a half-life of 6 days. **[3]**

10. **(a)** What is nuclear fission?

(b) How does the energy produced by nuclear fission compare with that produced in a chemical reaction? **[2]**

Chapter

2 Energy

The following topics are covered in this chapter:

- The effect of heating
- Heat transfer
- Energy resources
- How the electricity supply works
- Technology for the future

2.1 The effect of heating

Temperature and heat

OCR B P1a

Temperature is a measure of how hot an object is. The **temperature scale** we use measures temperature in **degrees Celsius (°C)**. On this scale 0°C is defined as the temperature at which pure ice melts – but temperatures can be much lower than this. Different temperatures can be shown on a **thermogram** – each colour represents a different temperature.

If the temperature changes *to* 1°C it is just above the freezing point of water – this is not the same as *by* 1°C. Don't confuse temperatures with temperature changes.

Heat is a form of energy. If an object is cooled until all the particles stop moving then they cannot lose any more kinetic energy. The amount of heat in the object is a minimum.

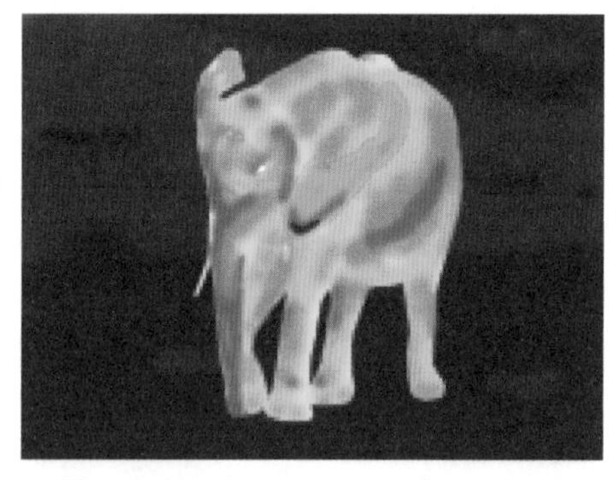

Fig. 2.1 A thermogram of an elephant.

Specific heat capacity

OCR B P1a

If the temperature of a solid, liquid or gas changes, then it has gained, or lost, energy. The amount of energy depends on:

- the temperature change
- the mass of the object
- the material the object is made from.

The **specific heat capacity** of a material is a measure of the energy of the material. It is different for different materials and tells us how much energy (in joules) you need to raise the temperature of one kilogram of the material by one degree Celsius ('specific' means 'for each kilogram').

The specific heat capacity of a material is the energy needed to increase the temperature of 1 kg of the material by 1°C:
Energy = mass × specific heat capacity × temperature change
Specific heat capacity is measured in J/kg °C.

Latent heat

OCR B P1a
AQA P1.11.5

Heating an object raises its temperature except at the **melting point** and **boiling point**. At these temperatures, the energy is being used to **change the state,** from solid to liquid, or liquid to gas. Because the temperature does not change, the heat given to the substance is called **latent heat** ('latent' means hidden).

Solid objects are held together by forces between the particles (atoms or molecules), and have a regular shape. The particles vibrate more as the object is heated. **Liquid** particles have enough energy to break the inter-molecular bonds and slide over each other. At the **melting point**, heating the solid does not increase the vibrations, but gives the particles enough energy to break the bonds. When a liquid **freezes**, it loses this energy to its surroundings.

A sketch of particles must show that the particle size doesn't change. Solid particles are regularly spaced and touching.

The liquid particles are still touching – there are no gaps large enough for another particle to fit in.

Gas particles are very widely spaced, so do not draw too many.

Gas particles have enough energy to separate completely. At the **boiling point**, heating the liquid breaks the inter-molecular bonds completely and the particles form a gas.

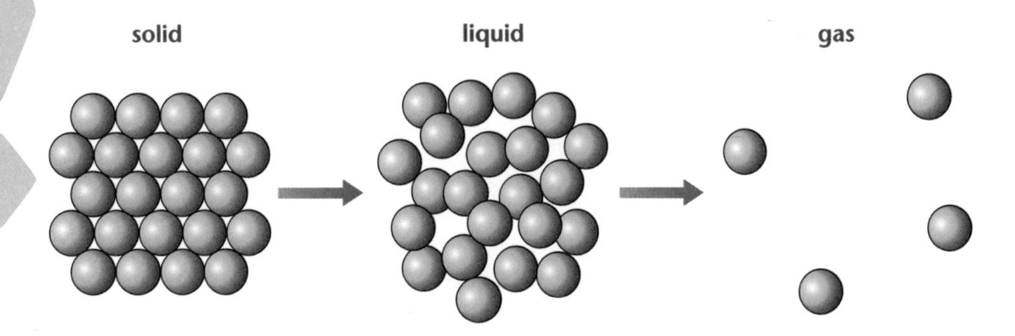

Fig. 2.2 There is a small increase in kinetic (movement) energy of the particles when an object changes from solid to liquid and a bigger increase when a liquid changes to gas.

> **KEY POINT**
>
> **The specific latent heat of melting of a material is the energy in joules needed to melt 1 kg of the material without changing its temperature.**
>
> **The specific latent heat of boiling of a material is the energy in joules needed to boil 1 kg of the material without changing its temperature.**
>
> **Specific latent heat is measured in J/kg.**

The specific latent heat of freezing is the same as that for melting, but the energy is given out by the material.

2.2 Heat transfer

Heating and cooling

OCR B P1c
AQA P1.11.1,1.11.5

Objects that are hotter than their surroundings cool down, objects that are colder than the surroundings heat up. The bigger the temperature difference between an object and its surroundings, the faster this happens.

Heat transfer by conduction

In a hot solid, the atoms vibrate more than in a cold one. They collide with atoms next to them and set them vibrating more. The kinetic energy is transferred from atom to atom. Metals are the best **conductors**, followed by other solids. Liquids are generally poor conductors. Gases are very poor conductors. Poor conductors are called **insulators.**

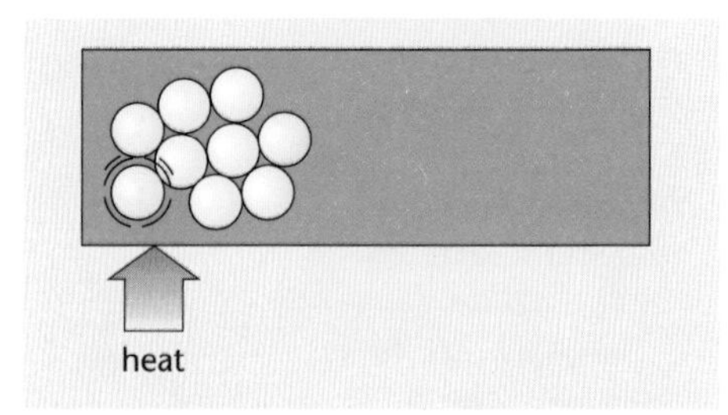

Fig. 2.3 Conduction in a solid. Energy is transferred from molecule to molecule.

Metals are good conductors because they have 'free' electrons that transport energy from the hot to the cold end of the material much faster.

Heat transfer by convection

Remember: hot gases and liquids rise, not 'heat'.

In a hot fluid (a gas or a liquid), the atoms have more kinetic energy than in a cold fluid, so they move more. They spread out and the fluid becomes less dense. The hot fluid rises above the denser cold fluid forming a **convection current**.

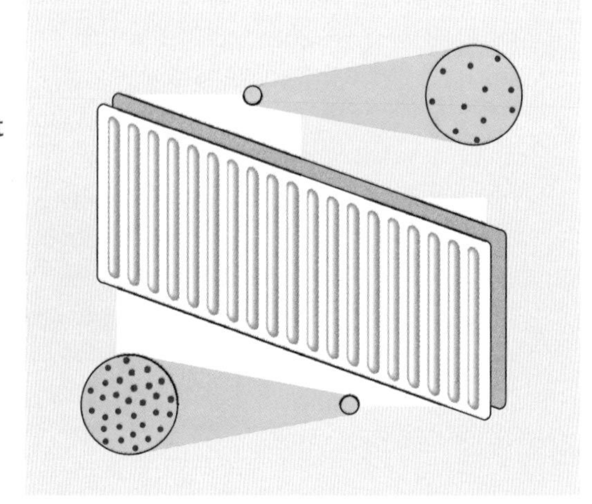

Fig. 2.4 The particles in the warm air are more widely spaced than those in the cold air, so the air is less dense.

Heat transfer by radiation

All objects **emit** (give out) and **absorb** (take in) **thermal radiation**. Some objects also **reflect** radiation. This radiation transfers energy in the form of infra-red electromagnetic waves. The hotter a body is the more energy it radiates. Radiation will travel through a vacuum – it does not need a **medium** (material) to pass through.

Dark and **matt** surfaces are **good absorbers** and **emitters** of radiation. **Light** and **shiny** surfaces are **poor absorbers** and **emitters** of radiation.

Insulation

In the winter we keep our homes much warmer than the outside temperature. This means that heat will be lost to the outside. If we reduce the heat lost, we use less fuel and it costs less. We can do this by **insulating** our homes.

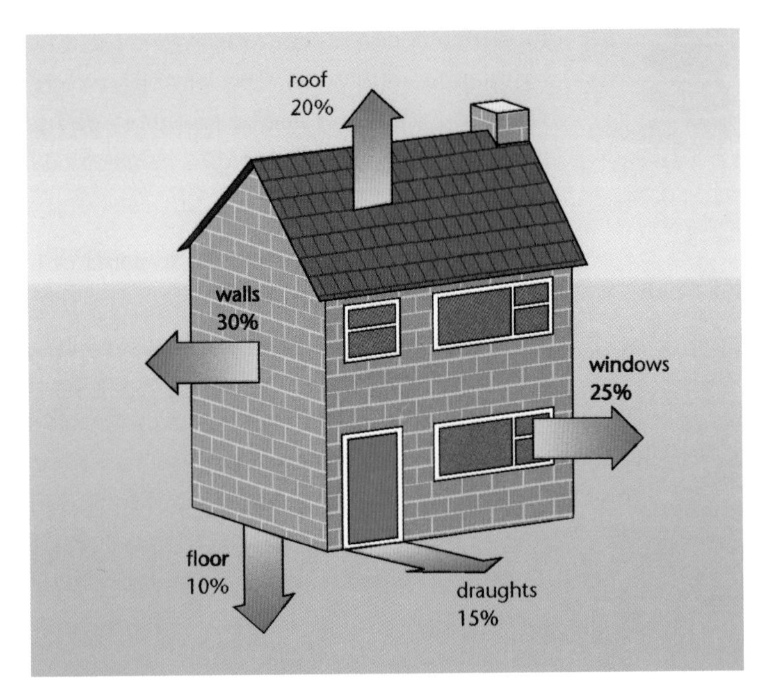

Fig. 2.5 Energy flow from an uninsulated house.

- **Still air** is a very good **insulator**. Some houses have **cavity walls**. The air gap between the two walls stops conduction. But air transfers heat if convection currents are set up. It is important to keep the air still. This can be done by **cavity wall insulation** – filling the cavity with a material containing trapped air, for example, foam or mineral wool.
- A lot of energy is lost through the roof, because **convection currents** are set up. **Loft insulation** uses fibreglass or mineral wool to keep the floor of the loft from getting hot.
- **Reflective foil** on walls reflects infra-red radiation.
- **Draught-proofing** stops the hot air leaving and cold air entering the house.

All these improvements cost money to buy and install, but they save money on fuel costs. You can work out the **payback time**, which is the time it takes before the money spent on improvements is balanced by the fuel savings, and you begin to save money.

KEY POINT

$$\text{Payback time (in years)} = \frac{\text{cost of insulation}}{\text{cost of fuel saved each year}}$$

If the price of the fuel increases the payback time will be less.

2.3 Energy resources

Energy questions

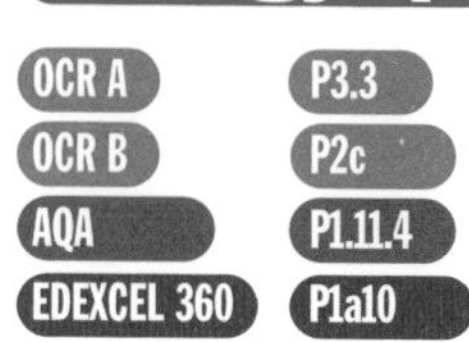

As members of a modern society we use a lot of **energy resources**. Some of these are used directly, especially oil – which is used as petrol and diesel for vehicles. A lot of our energy resources are used to make **electricity**, which we then use as a source of energy – so electricity is called a **secondary energy source**.

Readily available energy has the **benefit** of making our lives healthier and longer. The **drawbacks** are the amount of damage to the environment – the **pollution**. When we choose a fuel, we need to consider the effect on the environment. Some waste is toxic. **Carbon dioxide** is a waste product from using many fuels. Increased levels of carbon dioxide in the atmosphere cause some **global warming**. There is a lot of concern about climate change caused by global warming, so some countries, including the UK, are aiming to cut their emissions of carbon dioxide. The **risks** of using some energy resources are higher than others, but the risks to people because of extended power cuts are also high – for example; failure of hospital equipment, no heating in the winter and no refrigeration in the summer.

For more about carbon dioxide and global warming, see 4.6 The atmosphere and global warming.

Efficiency of energy transfer

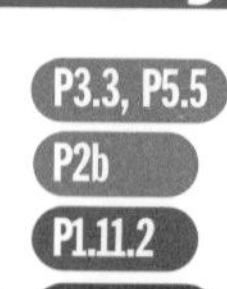

In a coal-fired power station for every 100 J of chemical energy stored in the coal that is burned, only 40 J is transferred to electrical energy. **Energy** is a **conserved** quantity, which means that the total amount of energy remains the same. It cannot be created. The 60 J of chemical energy that is not turned into electrical energy is wasted, and ends up as heat in the surrounding environment. This will cause a slight **temperature increase** in the surroundings. Energy becomes increasingly spread out and more difficult to use for further energy transformations.

Energy transfers can be shown on a **Sankey energy flow diagram**.

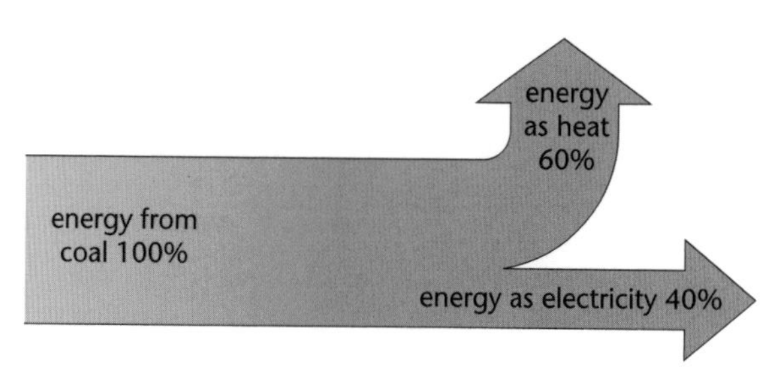

Fig. 2.6 Energy transformed in a coal-fired power station.

The width of the arrows is proportional to the amount of energy represented by the arrow.

Modern gas-fired power stations are more **efficient** than coal-fired power stations. More of the chemical energy stored in the gas is transferred to electrical energy.

KEY POINT

The efficiency of an energy transfer is the fraction, or percentage, of the energy input that is transferred to useful energy output:

$$\text{Efficiency} = \frac{\text{useful energy output}}{\text{total energy input}}$$

Or as a percentage:

$$\text{Efficiency} = \frac{\text{useful energy output}}{\text{total energy input}} \times 100\%$$

Gas-fired power stations are about 50% efficient whereas coal-fired power stations are about 40% efficient.

Generating electricity

Turning a **generator** produces electricity. To turn the generators we connect them to **turbines** and we use all the different energy resources available to turn the turbines. **Wind** and **water flow** can turn turbines directly. **Steam** is often used, produced by **heating water**. The heating is done by burning fuels, or using other heat sources. The diagram shows the parts of a coal-fired power station. In a **modern gas-fired** power station, the **hot exhaust gases** from the burners are used to turn the turbines, and then to heat water to steam which turns the turbines.

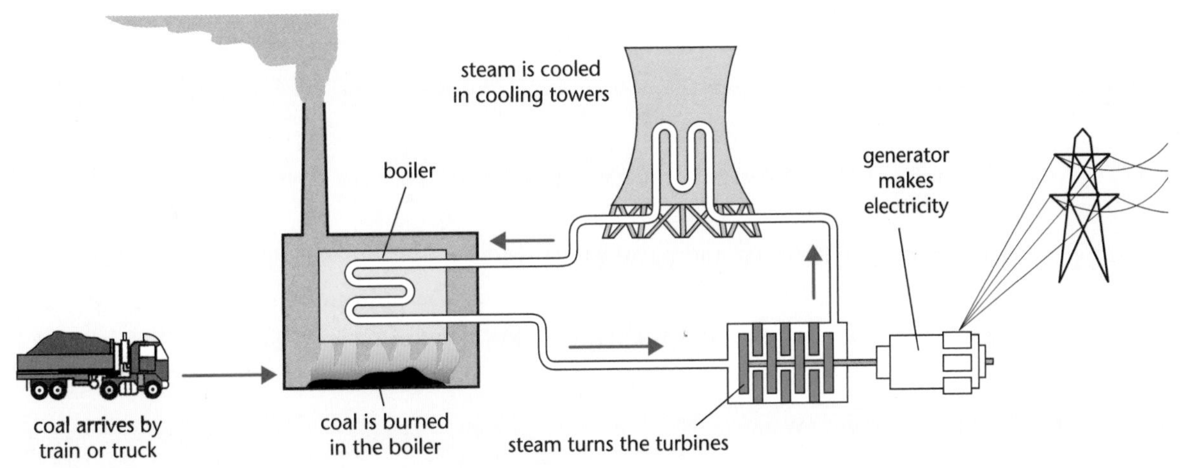

Fig. 2.7 A coal-fired power station.

The main fuels

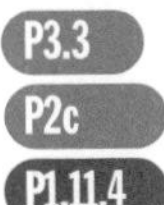

Electricity can be generated in large power stations from many different types of fuel:

- **Fossil fuels** – coal, natural gas and oil. These fuels were formed from the remains of **forests 300 million years ago**. None is being formed today so they will eventually run out. When they are burned, **sulfur dioxide** gas is formed, which dissolves in rain to form **acid rain**. Coal produces more sulfur dioxide than oil, and gas produces less. All the fossil fuels produce **carbon dioxide** when burned.

- **Nuclear** fuels – the nuclei of **plutonium** and **uranium.** Uranium is mined, but will not run out as quickly as fossil fuels. Plutonium is formed in nuclear reactors. If the plutonium is not processed into fuel, some of it can be used to make nuclear bombs. Energy is released when **nuclear fission** occurs – the nucleus splits in two. One advantage is that no carbon dioxide is formed. The radioactive materials produced remain dangerous to living things for millions of years (see 1.5 Nuclear waste). These include **pollution** from the fuel processing (both producing the fuel and making the waste fuel safe after it is used) and from the reactor when it reaches the end of its life. In addition there is the risk of an accidental emission of **radioactive** material while the power station is operating. These concerns mean that there are high maintenance costs and high decommissioning costs (the costs of taking the plant apart and making it safe at the end of its life).
- **Renewable** fuels. These are covered in the next section. They are fuels that are being made today and so will not be used up. Most of these are not used in large power stations. An exception is **hydroelectric** power, generated in some countries from the flow of large rivers and high waterfalls.

The table shows the fuels we use in the UK for generating electricity.

Fuel used in 2004	Percentage
coal	40
oil	<1
gas	34
nuclear	24
hydroelectric	<1
other renewables	<1
imports (electric cable to France)	<1

The UK oil and gas reserves under the North Sea are being used up. Since 2004, the UK has been a net importer of gas (it imports more than it exports). The UK is expected to become a net importer of oil in 2010.

The renewable fuels

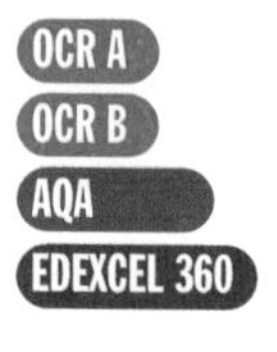

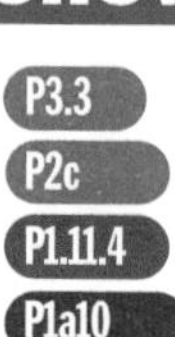

Most of the renewable fuels make use of the **Sun's energy**. It is the Sun that evaporates water and causes the rain to fill the rivers. It also causes convection currents that produce winds. The exceptions are **geothermal,** which results from radioactive decay inside the Earth, and **tides**, which are caused by the Moon.

- **Hydroelectric power (HEP)**. Fast-flowing water can be used to generate electricity. The UK does not have the fast-flowing large rivers needed to build large HEP stations, although there are some small ones in Scotland and Wales. There is no waste or pollution, but rainfall or snow is not constant so **dams** are needed. Building dams and flooding valleys and canyons changes the environment and causes conflict in some countries.

Fig. 2.8 A wind farm – visual pollution?

- **Wind turbines**. These transfer the kinetic energy of the air into electrical energy. In 2006 there were about 1500 wind turbines in the UK, and the number is growing. Some areas, particularly offshore, are windy all year round. Wind turbines can be made **rugged** (tough) enough to last in these conditions. They do not produce polluting waste gases, but some people consider them noisy and an eyesore – **visual pollution**. The wind is unpredictable and the amount of electricity generated will depend on the wind speed. Wind turbines take up a lot of space for the amount of electricity generated.
- **Solar cells**. These are not used to turn generators. They produce electricity inside the cell. This means there is no need for overhead power cables. The electricity is **direct current (d.c.)** – the direction of the current does not change. There is no polluting waste and no need to buy fuel. There are no moving parts so they are rugged and do not need much maintenance. They have a long life. They cannot produce power at night or in poor weather and are usually used to charge batteries. The big advantage is that they can be used in remote locations where there are no power lines. More details are given in the next section on solar energy.
- **Biomass** fuels, for example wood, straw, manure and household waste. These are products that are being formed today by plants or animals. Power stations need a steady supply. Suppliers need to be sure the power station will not close. The fuels can be burned directly, or fermented to produce **methane** gas. Using them will produce pollution – **carbon dioxide**, other gases and ashes. In the case of waste and manure, this pollution would be produced even if they were not used as fuel.
- **Geothermal**. There is a lot of heat below the Earth's crust. In places where the crust is thin and the heat is close to the surface, geothermal power stations can use this heat. Examples are in New Zealand and Iceland. There is no pollution, although the effect of extracting the heat may change the environment, and these areas experience earthquakes and volcanic action, which may damage a power station.
- **Tides**. In some places the change in the **height** of the water due to the tide is large enough to make it worth using to generate electricity. The tide does not depend on the weather. These areas are often important natural areas and holding back the water to run through turbines destroys the habitats of many birds and other animals.

Solar energy

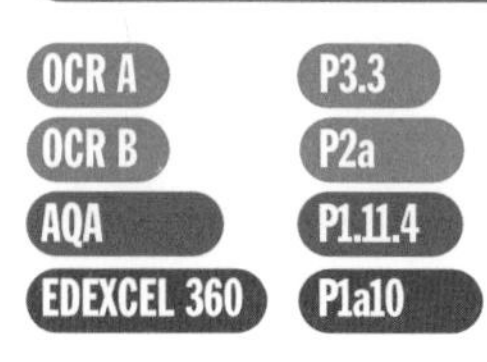

The Sun can be used as an energy resource without using generators. To collect the most energy from the Sun the collector must track the path of the Sun. Some installations do this, but some are set at the best angle and collect fewer of the Sun's rays. There are three ways of using the Sun's energy:

- **Passive solar heating** for buildings. Glass is transparent to light and short wave infrared radiation, but reflects longer wave infrared radiation. When the Sun shines on glass windows the light and short wave infrared radiation will pass through and warm the objects inside. The warm objects emit longer-wave infrared radiation, but this cannot escape through the glass. This effect is called the **greenhouse effect** because it explains how plants are kept warm inside a glasshouse. The effect can be used in **solar panels** to warm water as shown in Fig. 2.9. The water can be circulated in pipes around the house – the circulating water is often used to heat the water in the hot-water tank.

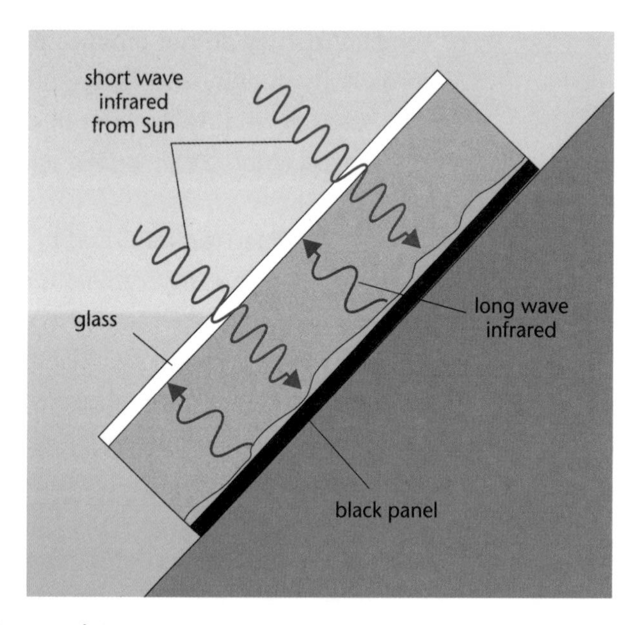

Fig. 2.9 A solar panel.

- **Solar furnace**. Light and infrared radiation is reflected from shiny surfaces. A curved mirror can be used to focus all the Sun's rays to a point. This can be used as a solar cooker to cook food.
- **Solar cells**. These are also called **photocells** or **photovoltaic cells**. They contain a crystal of **silicon**. Light falls on the crystal and gives **electrons** energy so they are released. The electrons flow as an **electric current**. The current can be increased by increasing the amount of light energy. This is done by increasing the surface area that the light falls on, or by increasing the light intensity. Solar cells are expensive to manufacture and are only about 30% efficient (although new developments may increase this to 50%). This is why they are not yet in widespread use.

2.4 How the electricity supply works

How generators work

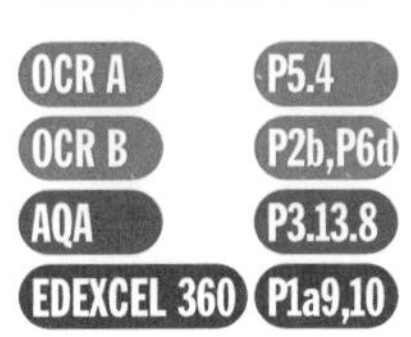

The diagram shows how a voltage is **induced** in a coil of wire by **moving** a **magnet** into or out of a **coil**. Moving the coil instead of the magnet would have the same effect. This effect is used in **dynamos** and **generators**.

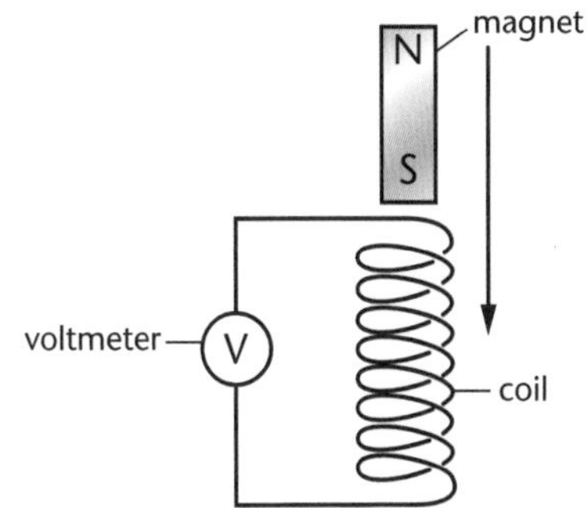

Fig. 2.10 A voltage is induced.

> **KEY POINT**
> **The dynamo effect occurs when a voltage is induced by:**
> - **moving a magnet near a coil**
> - **moving a coil near a magnet.**

There are three ways to increase the induced **voltage** (and get greater induced **current**):

- use stronger magnets
- use more turns of wire in the coil
- move the magnet (or the coil) faster.

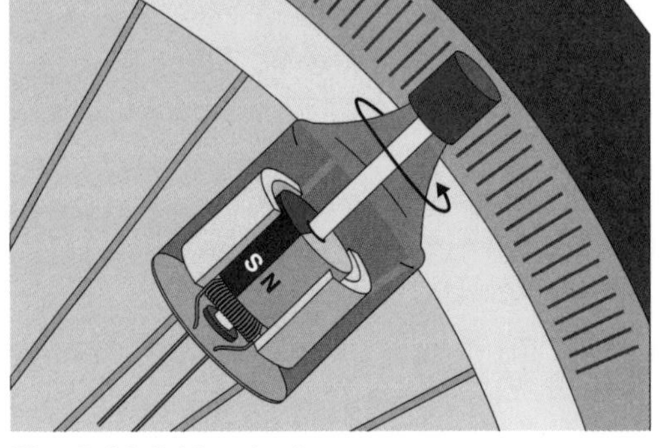

Fig. 2.11 A bicycle dynamo.

For Edexcel you need to know about the electric motor – see section 13.2.

The diagram shows a bicycle dynamo that uses a rotating magnet. As the magnet rotates faster, the induced current increases and the bicycle light gets brighter.

A generator in a power station uses an **electromagnet** to produce a **magnetic field**. The electromagnet rotates inside **coils of wire** so that the coil is in a **changing magnetic field** and a voltage is induced.

Alternating current (a.c.) and direct current (d.c.)

OCR A P5.4
OCR B P2b,6d
AQA P2.12.7
EDEXCEL 360 P1a9

For AQA, see section 8.2 Sound waves, for how to work out the period or frequency from an oscilloscope display.

The size of d.c. can change but it is always in the same direction. For a.c. a graph of the voltage alternates in the same way as a graph of the current.

Changing the direction of the **magnetic field** or the **movement** induces a voltage in the *opposite* direction. As the magnet rotates, the north pole and south pole swap over once in each complete rotation. This means the direction of the voltage and the current changes. This is called **alternating current (a.c.)**.

KEY POINT **Dynamos and a.c. generators produce alternating current (a.c.). Batteries and solar cells produce direct current (d.c.).**

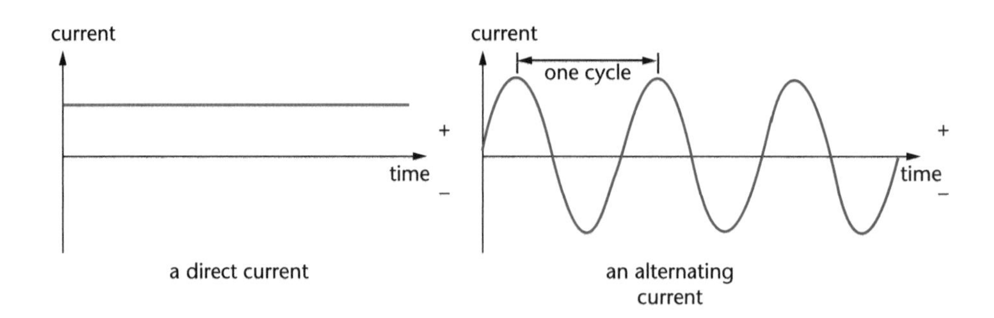

Fig. 2.12 d.c. and a.c.

In the UK, the a.c. generators at power stations that supply our mains electricity rotate **50 times in one second**. This means that there are 50 complete cycles each second. The number of cycles per second is called the **frequency**. Frequency is measured in cycles per second or in hertz (Hz), where 1 Hz = 1 cycle per second.

KEY POINT **in the UK, the mains frequency is 50 Hz, which is 50 cycles per second.**

Electrical energy and power

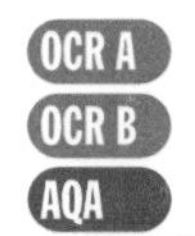
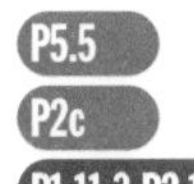
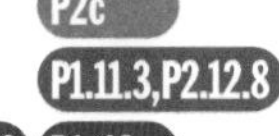

OCR A P5.5
OCR B P2c
AQA P1.11.3,P2.12.8
EDEXCEL 360 P1a10

Sound is made by the movement of a loudspeaker cone.

We use electrical appliances at home to transfer energy from the mains supply to:

- heating
- light
- movement (including sound).

In two hours, an electric lamp transforms twice as much energy as it transforms in one hour. The **power** of an electrical appliance indicates how much **electrical energy** it transfers in one **second** – in other words, the rate at which it transfers electrical energy into other forms of energy.

KEY POINT **Power is measured in watts (W) where 1 W = 1 J/s.**

Appliances used for heating have a much higher power rating than those used to produce light or sound.

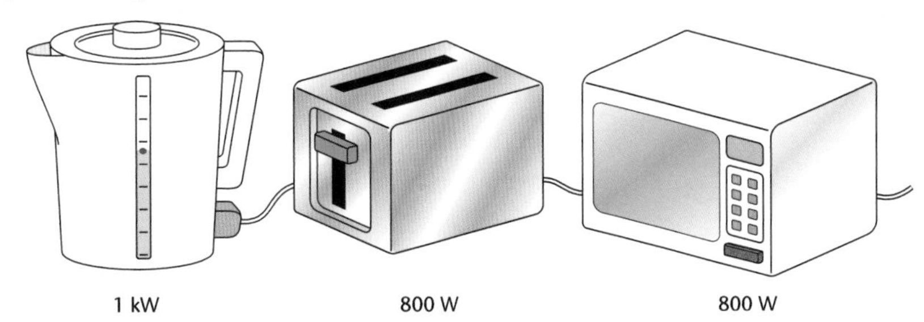

Fig. 2.13 Power ratings of electrical appliances.

The kilowatt-hour is a unit of energy – not power. (Power is measured in watts or kilowatts.)

The amount of **energy transferred** from the mains appliance depends on the **power** rating of the appliance and the **time** for which it is switched on. Energy is measured in joules, but electricity suppliers sell us electrical energy in **units** called **kilowatt-hours**. Electricity meters measure the energy transferred in kilowatt-hours.

> **KEY POINT**
>
> **Electrical energy is calculated by**
>
> **Energy = power × time**
>
> $$E = Pt$$
>
> **There are two sets of units used for energy:**
>
> - **The energy is in joules (J) when the power is in watts and the time is in seconds.**
> - **The energy is in kilowatt-hours (kW h) when the power is in kilowatts and the time is in hours.**

The cost of each unit of electrical energy – which is one kilowatt-hour of electrical energy – varies. At the moment it is about 10p. The electrical energy bill is calculated by working out the number of units used and multiplying by the cost of a unit.

> **KEY POINT**
>
> **Cost of electrical energy used is calculated from:**
>
> **Cost = power in kW × time in hours × cost of one unit**
>
> **or**
>
> **Cost = number of kW h used × cost of one unit**

For example:
For the 800 W microwave oven in Fig.2.13:
if it is used for half an hour and the cost of a unit is 10p:

$$\text{Cost} = 0.8\,\text{kW} \times 0.5\ \text{hours} \times 10\text{p/kW h}$$
$$\text{Cost} = 4\text{p}$$

Power, current and voltage

OCR A P5.5
OCR B P2c

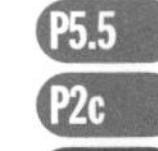
AQA P2.12.8

EDEXCEL 360 P1a10

For how to re-arrange the formula see page 225.

> **KEY POINT**
>
> **Electrical power is worked out using:**
>
> **Power = current × voltage**
>
> $$P = IV$$
>
> **Power is measured in watts (W), current in amperes (A) and voltage in volts (V).**

Transformers and the National Grid

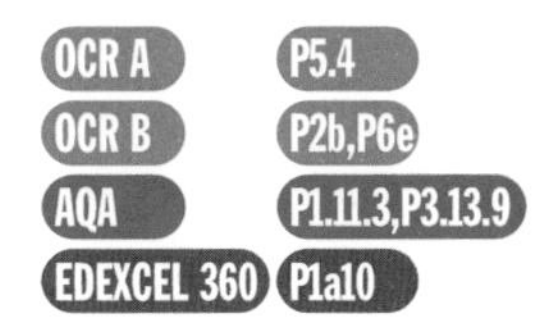

Transformers

A **transformer** changes the size of an **alternating voltage**. The voltage must be changing or the transformer will not work. One of the reasons we use an a.c. mains supply is so that we can change voltage using transformers.

Step-up transformers increase voltage, and **step-down transformers** decrease voltage.

The National Grid

The National Grid is the network of suppliers of electricity – the power stations and users of electricity – homes and workplaces. They are all connected together by power lines, some overhead and some underground.

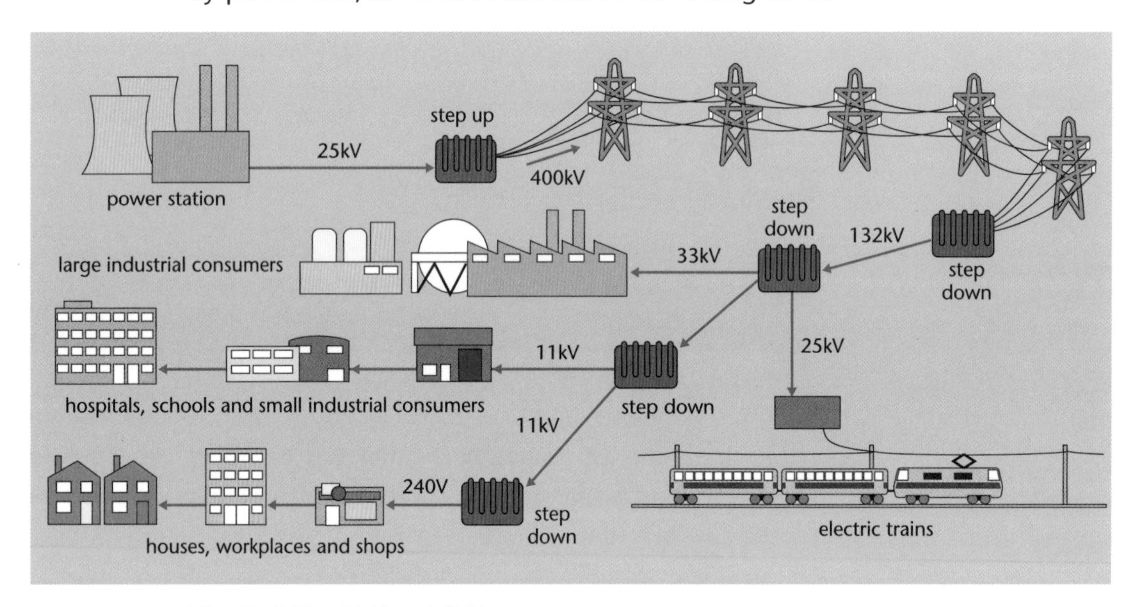

Fig. 2.14 The National Grid.

A National Grid has the following *advantages*:

- Power stations can be built where the fuel reserves are, or near the sea or rivers for cooling.
- Pollution can be kept away from cities.
- Power can be diverted to where it is needed, if there is high demand or a breakdown.
- Surplus power can be used to pump water up into reservoirs to be used to generate hydroelectric power when there is a peak in demand. (Dinorwig in Wales is a pumped storage power station.)
- Very large power stations can be built which are more efficient.

A National Grid has the following *disadvantages*:

- Power is wasted heating the power cables.
- Overhead power cables are an eyesore.
- Smaller generating projects such as wind turbines and panels of solar cells have difficulty competing with large suppliers.

We reduce the resistance of the cable by using thick copper, but the advantage of lower resistance has to be balanced against increased cost of cables and supports for the heavier cables.

Reducing power loss in power lines

To supply 100 kW of power through overhead power cables we could transmit 1 A at 100 kV or 10 A at 10 kV. (Using $P = IV$ the power is $P = 1\text{A} \times 100\text{V} = 100\text{kW}$ or $P = 10\text{A} \times 10\text{V} = 100\text{kW}$.)

The power cables have a **resistance** to the flow of current. The heating effect in the cables depends on the **resistance** and on the **current** in the cables. By making the current as small as possible we can reduce the energy wasted as heat in the cables. The current can be small if the voltage is large.

When supplying power through cables, a large voltage allows us to use a small current and this reduces energy waste by reducing the heating of the cables.

Mains voltage on specifications:
OCRA 230V
OCRB not specified
AQA about 230V
Edexcel 240V

Transformers are used to **step-up** the voltage. Several different voltages are used, for example the supergrid is at 400 kV as you can see in Fig. 2.14. **Step-down** transformers reduce this for us to use, for example, about 230 to 240 V in our homes.

2.5 Technology for the future

Battery technology

EDEXCEL 360 P1a9

As we make use of renewable energy, battery technology becomes more important for storing the energy. Electric cars and hybrid cars need reliable batteries (hybrid cars have an electric motor, and a petrol or diesel engine).

The **dry-cell** is a battery that contains a paste rather than a liquid, for example, the common zinc-carbon battery that you can buy for torches or toys. (An example of a battery that is not a dry-cell is a car battery - it contains liquid sulphuric acid.) One cell has a voltage of 1.5V. Batteries like a PP3, with a voltage of 9V are made of six 1.5V cells put together.

A battery has a **capacity** marked in **amp-hours**. A capacity of 1 amp-hour means the battery can deliver a current of 1 amp for 1 hour before it is flat. This means it could be used to deliver 2 amps for half an hour, or 100 milliamps for 10 hours.

KEY POINT

The battery capacity in amp-hours is:

The current it is supplying × the total time before it goes flat

Alkaline batteries have a similar construction but use different materials – zinc powder and manganese dioxide. The use of powder allows a faster electron flow so these batteries are used for higher currents, and they have a longer shelf life.

Re-chargeable batteries can be recharged many times. There are several types. One dry-cell type is the nickel-cadmium battery, which has the longest time before recharging, but cannot give such a high current as the others. Older batteries needed to be regularly discharged and recharged completely, or they lost the ability to be discharged and recharged completely (called a 'memory effect') – but newer batteries are better.

Cadmium is very toxic to all life forms – so there are environmental problems with battery disposal.

Another dry-cell type is the **nickel-metal hydride** battery. This type of battery does not suffer from the 'memory effect' and can have two or three times the capacity of the same size nickel-cadmium battery. It is also much less damaging to the environment. The Toyota Prius hybrid car uses nickel-metal hydride batteries.

Computers

The circuits that process all the information in a personal computer (P.C.) are contained in the microprocessor. This is an integrated circuit – a small chip that contains all the components in the circuit without the need for connecting wires. Keeping everything small and very close together allows the processes to take place very quickly.

The important components in a microprocessor are the transistors. The table shows how the number of transistors and the processor speed have changed since the first commercial microprocessors became available in the early 1970s. The microprocessors have not changed much in size since then – but the circuits have got smaller so that more and more can be packed into one integrated circuit.

Year	Number of transistors	Processor speed
1972	3500	200 kHz
1974	6000	2 MHz
1985	275 000	20 MHz
1997	7.5 million	200 MHz
1999	44 million	1.2 GHz

Processor speed is a measure of how fast the operations are done – A GHz processor carries out operations a thousand times faster than a MHz processor, which carries out operations a thousand times faster than a kHz processor.

Maglev trains

EDEXCEL 360 P1a9

Maglev is short for **magnetic levitation**. A maglev train has no wheels, it is suspended above the track by an electromagnetic force. The first commercial maglev train ran between Birmingham airport and the main railway station – a short distance of about half a kilometre. It was opened in 1984, but closed in 1995. There is now a line running for about 30 km between Shanghai and its airport, in China, and another line of about 18 km in Japan.

In 2007, plans to build a maglev link between Edinburgh and Glasgow were dismissed because some people were concerned that the maglev is 'unproven technology', though others did not agree. Maglev trains can travel much faster than wheeled trains. As well as keeping the train suspended above the track, the electromagnetic force is used to keep the train moving along the track. This linear motor works by continually changing the magnetic field (by changing the electric current in coils) so that the train is continually attracted forwards.

Superconductors have zero electrical resistance, so they do not get hot and waste energy when current flows.

In 2005, a design using high temperature **superconducting magnets** was demonstrated. The original maglev train systems did not use these, and have not been very successful, but including new technology could make them successful in the future.

A high temperature superconductor still has to be kept very cold – at the temperature of liquid nitrogen (–196°C).

KEY POINT

The maglev train may prove to be an example of how a scientific design or idea has to wait for appropriate technology (for example, superconducting magnets) to become available, at a reasonable cost, to make it a success.

HOW SCIENCE WORKS

OCR A P3.4/2.5
OCR B P2c
AQA P1.11.4
EDEXCELL 360 P1a10,P2.12

Nuclear reactors

In the UK, a number of nuclear reactors are reaching the end of their useful life. Our electricity requirements are expected to increase in the future. How should we generate our electricity in the future? There is no 'correct' answer. The government consults people and interested groups about their views and then decides. If people feel very strongly about the decisions they organise protests.

Gemma

"Wind turbines don't cause pollution, and I don't think they are an eyesore – pylons and power stations look much worse."

"Nuclear power stations produce dangerous waste that will be radioactive for millions of years – it is not fair to leave that for our descendents. Suppose there is an accident? The whole country may be dangerously radioactive."

Joanna

David

"The world could survive a nuclear power station blowing up – but global warming could make it uninhabitable – like Venus. We must stop burning fossil fuels."

"The whole country would have to be covered with windmills to generate enough electricity. There will be no power when the wind drops – or in gales when it is too windy to use wind turbines safely."

Ann

Daley

"We should think small – lots of solar cells and small wind turbines. Battery technology is improving all the time so we store the energy for when there is no sunshine or wind."

"It's crazy to say we won't build nuclear power stations and then buy electricity from France – generated in their nuclear power stations. Modern nuclear power station designs are much safer than the old ones."

Jose

1. Write down the name of a person who is talking about:
   (a) a risk of using nuclear power stations
   (b) a risk of fosssil fuels
   (c) a risk of relying on wind turbines for electricity
   (d) a benefit of using wind turbines for electricity **[4]**
2. Which of these questions can be answered by science?
   Q1 How much will the background activity increase if a nuclear power station is built?
   Q2 Which will spoil the view more, a nuclear power station or a wind farm?
   Q3 How many wind turbines are needed to produce the same amount of electricity as a nuclear power station?
   Q4 Which should be built at a particular site, a nuclear power station or a wind farm? **[4]**

Exam practice questions

1. Which of these improvements, to the insulation of a building, works by stopping convection:
   A Two walls with a gap between (cavity walls)
   B Two panes of glass with a gap between (double glazing)
   C Filling the wall cavity with foam (cavity wall insulation)
   D Silver foil wall lining behind radiators **[1]**

2. A power station burns coal and converts 3.50 MJ of chemical energy to 1.19 MJ of electrical energy every second. The efficiency of the power station is:
   A 2.31%
   B 29%
   C 34%
   D 47% **[1]**

3. When we 'use electricity' what do we measure in kilowatt-hours?
   A cost of electricity supplied
   B energy used
   C power output
   D time taken to use 1 kilowatt **[1]**

4. Which of the following devices transfers kinetic energy to electrical energy?
   A an a.c. generator
   B an electric motor
   C a turbine
   D a transformer **[1]**

5. Fill in the gaps. (Use words once, more than once, or not at all.) Choose words from:

   2100 4200 8400 0.5 1 2 more than less than the same as

   Water has a specific heat capacity of 4200 J/kg °C. This means that to raise the temperature of one kilogram of water by 2°C needs ____**1**____ J of energy. The specific heat capacity of aluminium is 880 J/kg °C. If one kilogram of aluminium is given 440 J of energy its temperature will rise by ____**2**____ °C. The heat needed to increase the temperature of a kilogram of water by 5°C is ____**3**____ the heat needed to increase the temperature of a kilogram of aluminium by 5°C. **[3]**

Exam practice questions

6. Fill in the gaps. (Use words once, more than once, or not at all.) Choose words from:

air gaps **cavity** **conduction** **convection** **convection currents** **radiation** **walls**

To improve the insulation of a building you need to reduce the heat lost by
_____1_____, _____2_____ and _____3_____.
Loft insulation works by reducing _____4_____ in the roof space.
cavity walls work because the _____5_____ reduce _____6_____ through the
_____7_____.
Putting foil behind radiators reflects _____8_____. **[8]**

7. Here are some statements about electricity generation in the UK. Write **T** for each **true** statement and **F** for each **false** statement.

(a) Most of the electricity in the UK is generated using fossil fuels.

(b) To reduce carbon dioxide emissions we must reduce our use of nuclear power stations.

(c) One large wind farm could replace several coal-fired power stations.

(d) Wind turbines are a renewable energy resource.

(e) Nuclear power stations produce radioactive waste that will remain radioactive for thousands of years.

(f) Natural gas fired power stations do not emit carbon dioxide. **[3]**

8. Put these statements in order to describe how a power station produces electricity.

A The steam is used to turn a turbine.
B Coal is burned in the furnace.
C This turns the generators to produce electricity.
D The water in the boiler is heated. **[3]**

9. A combined cycle gas turbine (CCGT) power station burns gas and uses the hot exhaust gases to turn the turbines. The hot exhaust gases are then used to heat water to steam to turn more turbines. Every second 755 MJ of electricity is generated from 1300 MJ of chemical energy in the gas.

(a) Use the equation

$$\frac{\text{electrical energy output}}{\text{fuel energy input}} \times 100\%$$

to calculate the efficiency of the power station. **[2]**

(b) Explain two ways a CCGT power station is different from a coal-fired power station. **[2]**

(c) Explain one disadvantage of using gas to fuel a power station. **[1]**

Exam practice questions

10. This table shows some electrical power ratings for wind generators.

Model	Maximum power rating of one turbine (kW)	Wind turbine recommended for:
A	0.025	battery charger
B	0.5	electricity for a caravan
C	5	domestic electricity for a house
D	25	electricity for a school
E	500	a wind farm
F	1500	an offshore wind farm

(a) What will affect the amount of power that a given wind turbine can generate? **[1]**

(b) What model is recommended for a house? **[1]**

(c) A house has an average power use of 0.5 kW. Describe how the actual power used might be **(i)** higher than average and **(ii)** lower than average **[2]**

(iii) Suggest two reasons model B is not recommended for the house. **[2]**

(d) Wind turbines are often used to recharge banks of batteries. Why is this a good idea? **[1]**

(e) A power station generates 750 MW of electrical power. How many wind turbines would be needed to replace it with **(i)** an offshore wind farm and **(ii)** an onshore wind farm **[2]**

(f) What is meant by 'the pay-back time for a turbine?' **[1]**

Exam practice questions

11. Ian decides to check his electricity bill:

Meter reading at start of quarter	487 612 kilowatt-hours
Meter reading and end of quarter	489 360 kilowatt-hours
Cost of one unit	10p

(a) How many units has Ian used? **[1]**

(b) How much is his electricity bill for this number of units? **[1]**

Ian decides to try to cut down the amount of electricity he is using so he keeps some notes for a week:

Appliance	**Power rating (kW)**	**Time used each day (hours)**
electric fire	3	2
electric kettle	1.2	0.5
electric lamp	0.1	5
computer	0.3	4
television	0.5	3

(c) How many units does the electric fire use in a day? **[1]**

(d) How much does it cost to use the fire for a day? **[1]**

(e) Which appliance is used for the most time? **[1]**

(f) How much does it cost to use this appliance for a day? **[1]**

Chapter

3 Waves

The following topics are covered in this chapter:

- *Describing waves*
- *Seismic waves*
- *Sound and ultrasound*
- *Electromagnetic waves*
- *Wave properties*

3.1 Describing waves

Types of wave

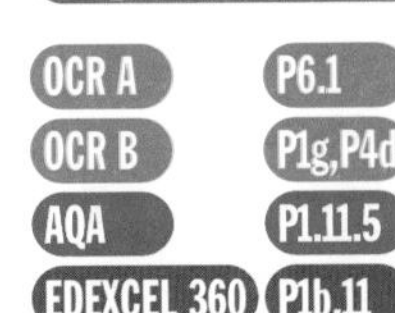

A **wave** is a **vibration** or disturbance which is transmitted through a material – called a **medium** – or through space. Waves transfer **energy** and can also be used to transfer **information** from one place to another, but they do not transfer material.

A common mistake is to mark the amplitude from the top of a peak to the bottom of a trough – this is twice the amplitude.

Transverse waves

A **transverse wave** has the vibrations at **right angles** (perpendicular) to the direction of wave travel. The wave has **peaks** (or **crests**) and **troughs**, as shown in the diagram. The **amplitude** is the maximum displacement (change in position) from the undisturbed position. The **wavelength** (symbol λ) is the distance between two neighbouring peaks or troughs.

For how to re-arrange the formula see page 225.

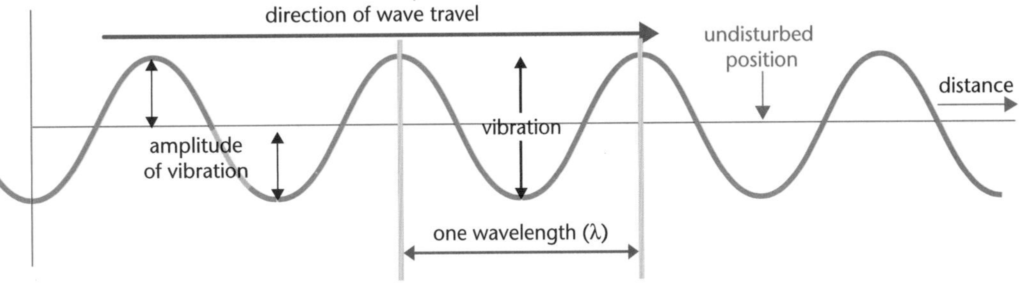

Fig. 3.1 A transverse wave.

The **frequency** is the number of complete waves that pass through a point in one second. It depends on how fast the source of the waves is vibrating. The frequency is measured in **hertz** (Hz) where one hertz is one cycle (wave) per second.

The **wave speed** depends on the medium that the wave is travelling through. As the frequency increases the wave speed does not change, but the wavelength will decrease. This is shown in Fig. 3.2.

a wave on a rope

a higher frequency wave

Fig. 3.2 Waves with different frequency and wavelength.

The **wave equation** relates the wavelength and frequency to the wave speed.

> KEY POINT
>
> **The wave equation, for all waves:**
>
> **wave speed = frequency × wavelength**
>
> $v = f\lambda$
>
> **If f is in Hz and λ is in m, then v is in m/s.**
>
> **The equation can also be written as $\lambda = v/f$ and $f = v/\lambda$.**

Examples of transverse waves are: water waves, light and other electromagnetic waves, and the seismic waves called S-waves.

Longitudinal waves

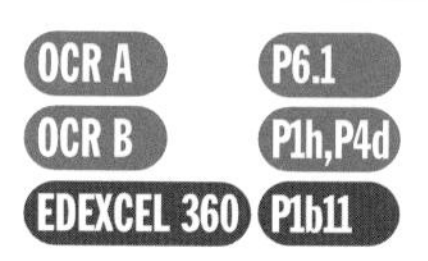

A **longitudinal wave** has the vibrations **parallel** to (along the same direction as) the direction of the wave travel. As shown in the diagram, the wave has **compressions** (or squashed parts) and between these are stretched parts called **rarefactions**.

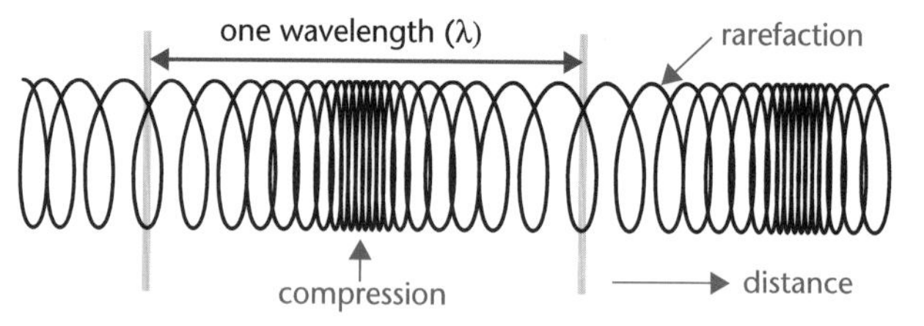

Fig. 3.3 A longitudinal wave.

One wavelength is the length of one complete wave – a compression and a rarefaction. Longitudinal waves show the same behaviour (for example reflection and refraction) as transverse waves.

3.2 Seismic waves

P-waves and S-waves

There are two types of shock waves called **seismic waves** that travel *through* the Earth (other types travel over the surface). These are called **P-waves** and **S-waves**. They can be caused by an earthquake, or a large explosion, and are detected at monitoring stations around the Earth by instruments called **seismometers.** Fig. 3.4 shows a **seismograph** – a record of the waves received at a monitoring station.

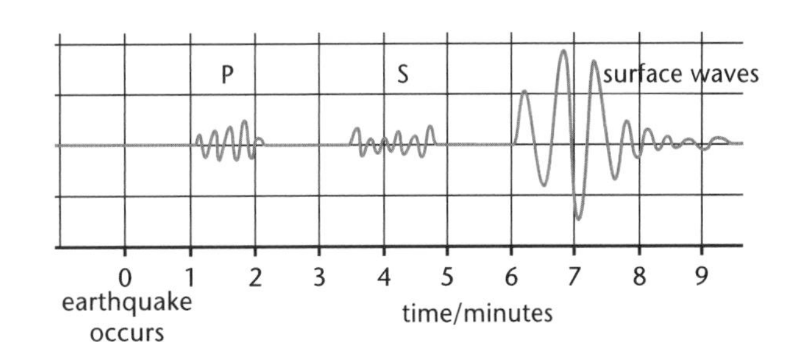

Fig. 3.4 A seismograph.

P-waves

P-waves, or primary waves, are **longitudinal** waves that travel through solid and liquid rock. They travel faster through the Earth than other seismic waves so they are the first to be detected after an earthquake.

Transverse waves can travel on the surface of liquids – but not through them.

S-waves

S-waves, or secondary waves, are **transverse** waves and can only travel through the solid materials in the Earth. They are detected after primary waves because they have a lower speed.

Fig. 3.5 shows the paths of P-waves and S-waves through the Earth following an earthquake. The point on the Earth's surface directly above the earthquake is called the **epicentre**. On the opposite side of the Earth to the epicentre no S-waves are detected. This tells us that there must be a part of the Earth that is liquid and only P-waves pass through it. This is how we know there is a **liquid outer core**.

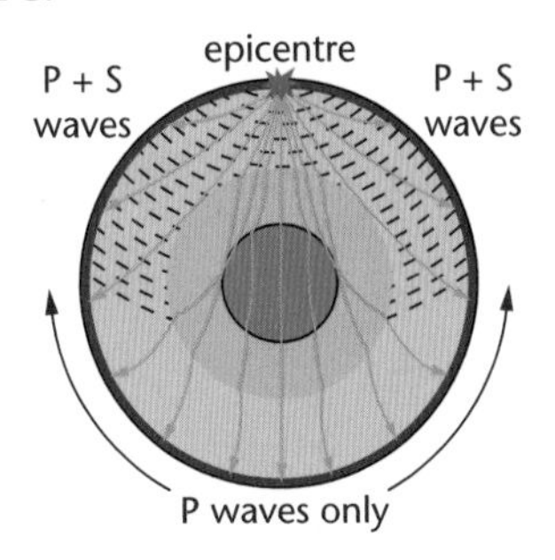

Fig. 3.5 P and S waves travel through the Earth.

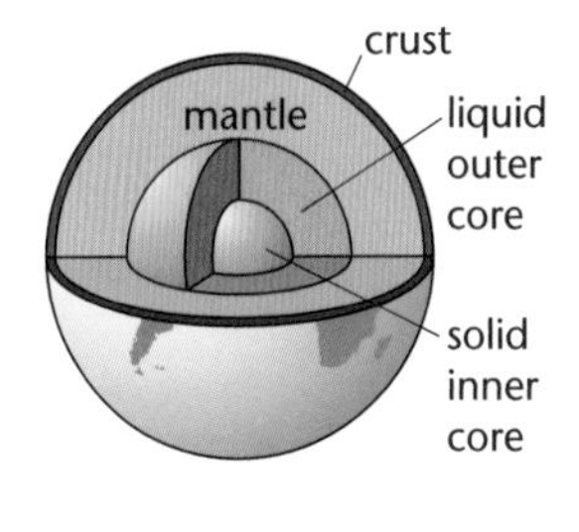

Fig. 3.6 The Earth's structure.

KEY POINT

The S-waves form a shadow region on the opposite side of the Earth to an earthquake. This shows that part of the Earth's core is liquid.

We can investigate the structure of the Earth's crust by setting up monitoring equipment at different points and setting off a controlled explosion. We record the waves arriving at the monitoring points. The time after the explosion that waves take to arrive depends on the speed of the waves (which in turn depends on the medium they pass through) and whether they have been reflected at a boundary between different materials. Analysing this data gives us information about the structure of the rocks.

3.3 Sound and ultrasound

Longitudinal waves

AQA P3.13.5,6
EDEXCEL 360 P1b11

Sound waves are **longitudinal** waves. They pass through solids, liquids and gases. **Ultrasound** waves are sound waves with a **frequency** that is too high for humans to hear. Like all waves, sound waves and ultrasound waves can be reflected – the reflections are called **echoes**. Echoes can be used for measuring distances. Fig. 3.7 shows how the depth of water can be measured by reflecting an ultrasound pulse off the seabed.

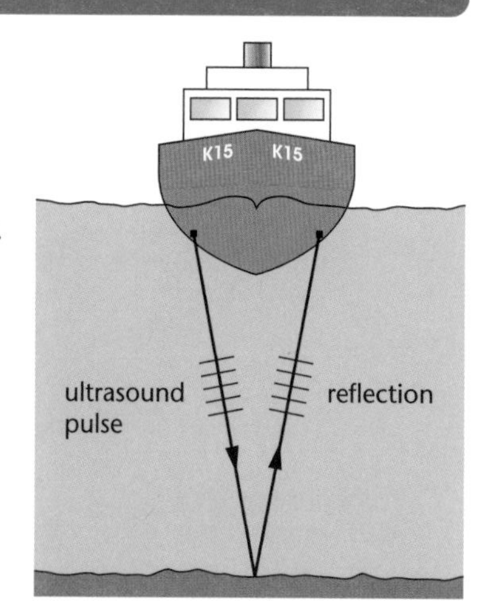

Fig. 3.7 Using echoes to measure distance.

Echo sounding works like this:

- An ultrasound pulse is emitted from a vibrating crystal.
- The same crystal detects the reflected pulse.
- The time for the pulse to travel from the crystal to the seabed and back to the crystal is recorded.
- The distance can be worked out using the equation speed = distance/time. The speed of the ultrasound waves in water is known, so distance = speed × time. The depth is half the distance the wave travelled, so depth = $\frac{\text{speed} \times \text{time}}{2}$

When you work out the depth, remember that the pulse has travelled twice the depth – there and back.

Ultrasound is also used to **scan** parts of the body, like the eye or an unborn foetus. This works because part of the pulse is reflected at each **boundary between** different **tissues** (for example skin and bone). The reflections from the tissue boundaries are all used to build up a picture. Ultrasound is much safer than X-rays because it does not damage body cells or DNA and does not cause mutations.

3.4 Electromagnetic waves

Transverse waves

OCR A P6.3
OCR B P1g

AQA P1.11.5
EDEXCEL 360 P1b11

Electromagnetic waves are transverse waves, which are made up of vibrating magnetic and electric fields. They can travel through a **vacuum** and all travel through space at a speed of **300 000 km/s**. The different types of electromagnetic waves form the **electromagnetic spectrum**, shown in Fig. 3.8.

frequency/Hz	10^5	10^8	10^{11}	10^{14}	10^{17}	10^{20}
	radio waves	microwaves	infra-red	light	ultraviolet	gamma rays / X-rays
wavelength/m	10^3	1	10^{-3}	10^{-6}	10^{-9}	10^{-12}

Fig. 3.8 The electromagnetic spectrum.

The spectrum of electromagnetic waves is continuous from the **longest** wavelengths (**radio waves**) through to the **shortest** wavelengths (**gamma rays**). These **wavelengths** are related to the **frequencies** using the wave equation – radio waves have the lowest frequency and gamma rays the greatest frequency.

The **higher** the **frequency** the more **energy** the waves have. The wavelength of visible light is about half a thousandth of a millimetre – so small that it is not obvious that it is a wave. In fact, sometimes light behaves as a stream of particles and sometimes as a wave.

3.5 Wave properties

Absorption, emission and reflection

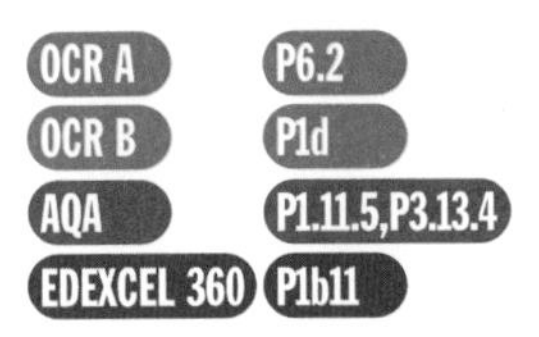

All surfaces **emit** electromagnetic radiation that depends on their **temperature**. They also **transmit**, **absorb** and/or **reflect** some of the radiation that falls on them. How much is transmitted, how much absorbed and how much reflected depends on the surface. Reflection happens to waves and particles. If a wave, or a ball, strikes a wall at an angle it will be reflected so that the **angle of incidence** is equal to the **angle of reflection**.

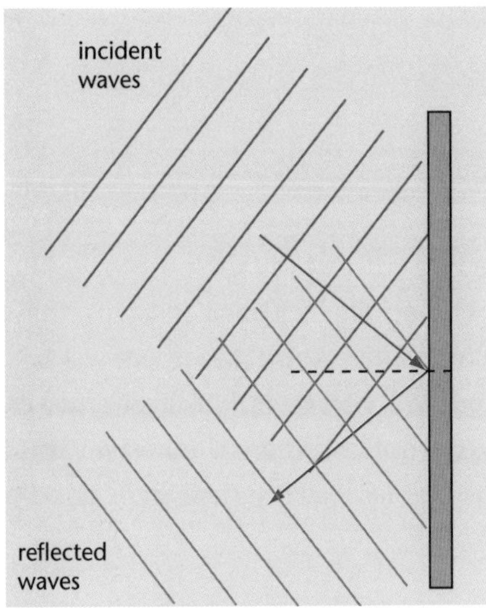

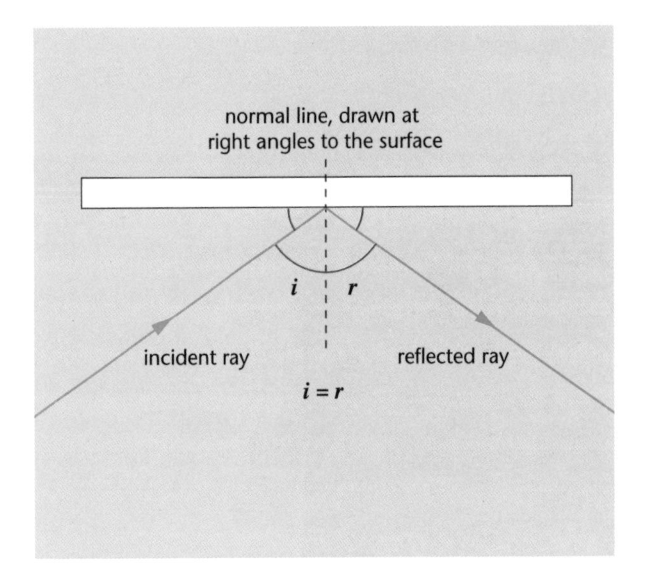

Fig. 3.9 Reflection.

Refraction

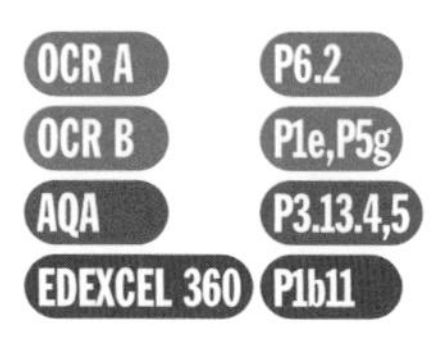

When waves enter a **denser medium**, they slow down. When they enter a less dense medium, they speed up. In both cases, this may cause them to **change direction**. This happens to water waves but also to particles, as you can show by rolling a ball at an angle down a ramp. When the slope of the ramp changes the direction of the ball changes.

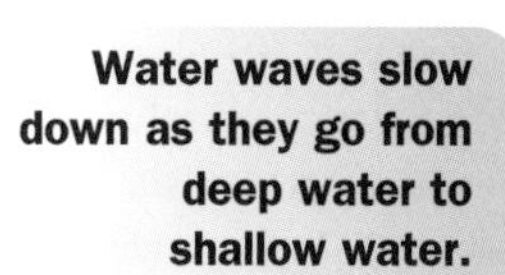

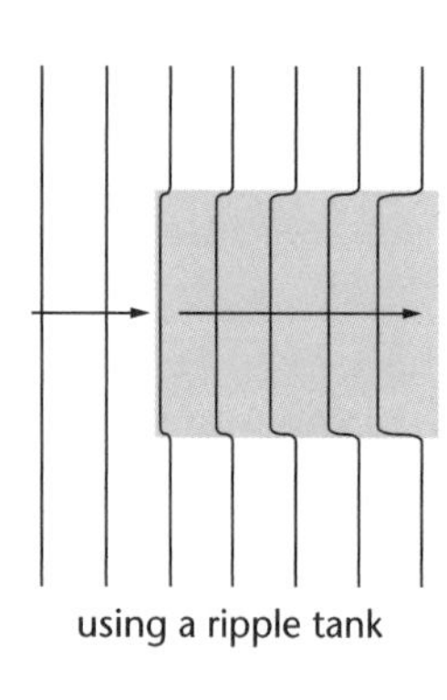

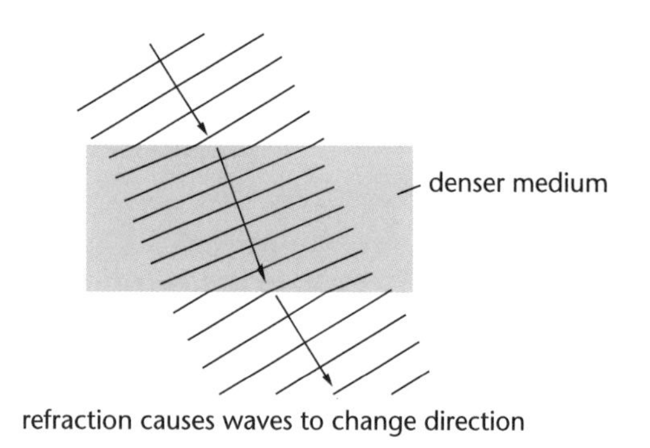

Fig. 3.10 Refraction.

Diffraction

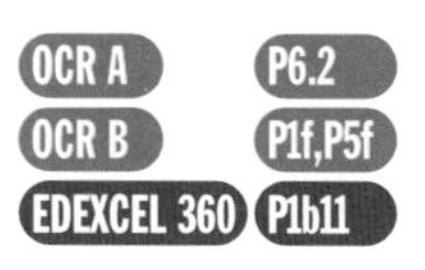

Diffraction is the **spreading** out of a wave when it passes through a gap. The effect is most noticeable when the gap is the same size as the wavelength. Particles cannot be diffracted. So diffraction is good evidence for a wave.

When answering questions it is important to say that the amount of spreading depends on the size of the gap compared with the wavelength

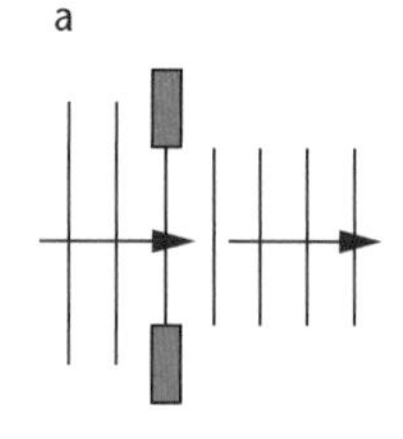

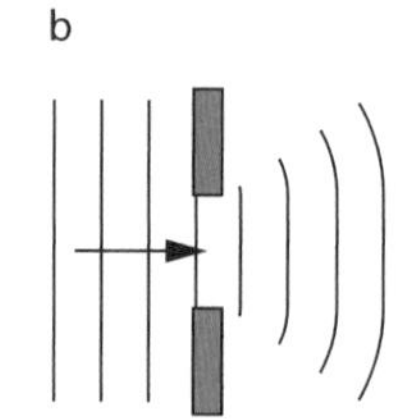

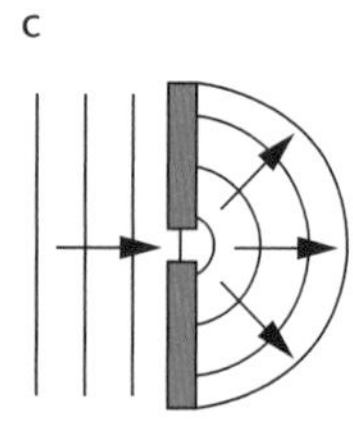

Fig. 3.11 Diffraction.

Interference

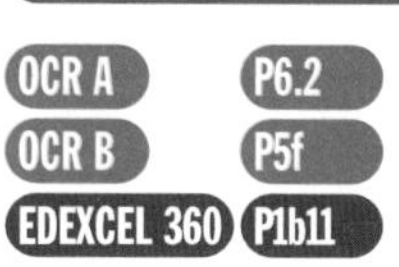

Interference occurs where two waves overlap. If the waves have the same **amplitude** and **wavelength** and are in **phase** (in step), they can **interfere**. Fig. 3.12 shows that if the crests arrive **together** there will be **constructive** interference and the amplitude will **increase**. If a crest arrives at the same time as a trough the two will **cancel** out and there is no wave. This is **destructive** interference. Particles cannot interfere, so interference is good evidence for a wave.

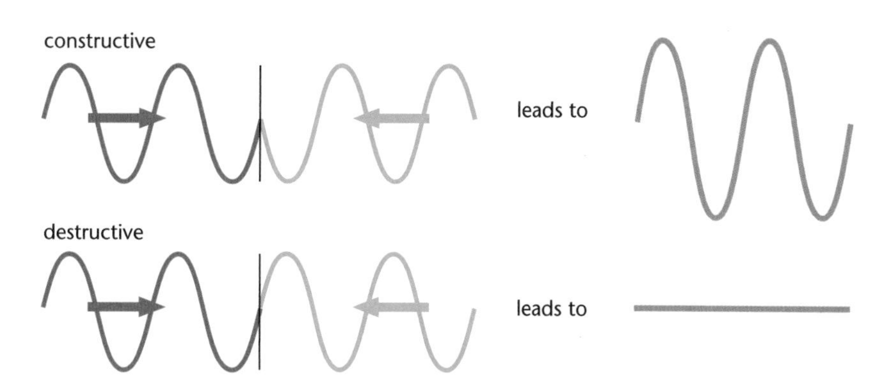

Fig. 3.12 Wave interference.

Total internal reflection

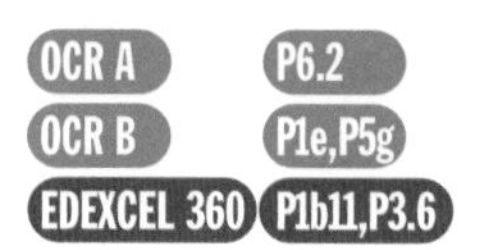

Total internal reflection can only happen when light travels from a **dense to a less dense medium** – for example from glass to air. If the angle of incidence is so large that the angle of refraction would be greater then 90° then it is impossible for the light to leave the glass – so total internal reflection occurs.

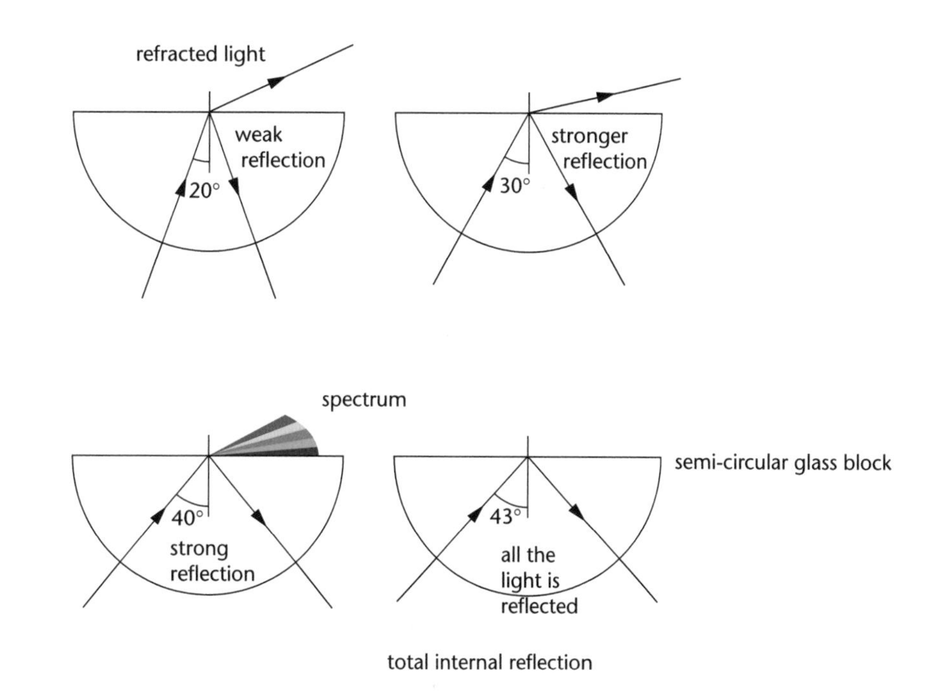

Fig. 3.13 When air meets an air–glass boundary.

Transparent materials have a **critical angle**. When the angle of incidence equals the critical angle, the angle of refraction is 90°. At angles greater than the critical angle, total internal reflection occurs. The critical angle for a glass–air boundary is about 42°. Total internal reflection can also occur at a Perspex–air boundary and at a water–air boundary.

HOW SCIENCE WORKS

OCR A P1.4
OCR B P1h
EDEXCEL 360 P1b.11

Can earthquakes be predicted?

Each year on Earth, there are about 19 large earthquakes of magnitude 7.0 or higher. (Magnitude is a scale for measuring earthquake size.) This is an average and the table shows how the number has ranged from 6 large earthquakes in 1986 to 34 in 1957.

Most large earthquakes occur along the fault zones bordering the Pacific Ocean, so we can predict that this is *where* large earthquakes will occur, but it is not so easy to predict *when*. Some of the faults are well known and predictions are made of where and when an earthquake will occur based on how much time there has been between large earthquakes in the past. This forecasting technique can be used only for well-understood faults, such as the San Andreas fault in California. It is not possible for many faults, for example those that caused the 1994 earthquake in Northridge, California, and the 1995 earthquake in Kobe, Japan.

One successful earthquake prediction was for the 1975 Haicheng earthquake in China, when an evacuation warning was issued the day before the earthquake. People were warned to evacuate because there was an increase in the small earthquakes that can happen before a large one. Unfortunately, most earthquakes do not have these small warning earthquakes, for example there was no warning of the 1976 Tangshan earthquake in China, which caused an estimated 250 000 fatalities.

Large earthquakes (magnitude greater than 7) each year since 1956:

1956	15	1982	10
1957	34	1983	15
1958	10	1984	8
1959	15	1985	15
1960	22	1986	6
1961	18	1987	11
1962	15	1988	8
1963	20	1989	7
1964	15	1990	18
1965	22	1991	16
1966	19	1992	13
1967	16	1993	12
1968	30	1994	13
1969	27	1995	20
1970	29	1996	15
1971	23	1997	16
1972	20	1998	12
1973	16	1999	18
1974	21	2000	15
1975	21	2001	16
1976	25	2002	13
1977	16	2003	15
1978	18	2004	16
1979	15	2005	11
1980	18	Total 1900–2005 = 2061 earthquakes	
1981	14	Average: 19.4 each year	

The data shows that there is no pattern to the number of large earthquakes. It tells us that we can expect about 19 or 20 each year, but there have been years when there were as few as 6 and as many as 34, so this could happen again.

HOW SCIENCE WORKS

Some people claim to sense when an earthquake is about to happen. Others say that animals know and behave strangely. Scientists have studied these effects and found that people are just as likely to predict earthquakes that don't happen, and that animals behave strangely when no earthquake occurs.

How can we prevent loss of life in earthquakes?

It is not earthquakes that kill people, but collapsing buildings. In the 1920s, the geologist Bailey Willis tried to persuade Californians to pass safe building regulations that would make buildings less likely to collapse in an earthquake. He thought that the authorities were acting too slowly and began to exaggerate, talking of a serious earthquake within 10 years. A building organisation showed that his claims were scientifically unsound, and because he was discredited Willis had to give up his work and the building regulations were rejected. Then in March 1933 there was an earthquake at Long Beach in California in which 15 school buildings collapsed and 40 were damaged. Only the fact that the earthquake occurred at 5.45 pm prevented the deaths of thousands of children. This time scientists campaigned for safer building regulations with success and the regulations were passed in May 1933.

This example shows that it is important to use scientifically correct data and facts to support your arguments – otherwise people can point out scientific errors and use them to discredit the argument. Safe building regulations were a good idea, whether an earthquake occurred in 10 years or 50 years, but by saying that an earthquake would occur in 10 years, Willis lost the argument.

Since those first building regulations, the designs have improved and when there is an earthquake today, the earthquake engineers investigate immediately to see what can be learned and how building techniques can be improved for the future. Another way to save lives is to use the delay between the P-waves (which arrive first) and the more damaging S-waves to give a warning. The Japanese bullet train, the Shinkansen, has sensors to automatically detect the P-waves and shut down, so that at least the train is slowing down before the S-waves arrive. Sensors can be used to shut down other processes in a similar way.

1. The average number of earthquakes was 19.4. How many of the 50 years in the table were average years, with 19 or 20 earthquakes? **[1]**
2. The range of the number of earthquakes per year was
   - **A** between 11 and 15
   - **B** between 0 and 19.4
   - **C** between 6 and 34
   - **D** between 1956 and 2005 **[1]**
3. Scientists look for a correlation between other phenomena and an earthquake, in the hope of predicting earthquakes.
   - **(a)** What correlation did they notice that helped them to predict the 1975 Haicheng earthquake?
   - **(b)** Give an example of a correlation people said they noticed but scientists disproved. **[2]**

Exam practice questions

1. Which word describes the number of waves passing a point in one second?

A amplitude
B frequency
C wavelength
D wave speed **[1]**

2. Earthquake waves known as P-waves can pass through

A liquids only
B solids and liquids
C solids only
D neither solids nor liquids **[1]**

3. The diagram shows how water waves spread out after passing through a gap. This effect is called

A diffraction
B dispersion
C reflection
D refraction **[1]**

4. Use some of the words below to complete the sentences.

disturbance energy longitudinal matter medium transverse vibrates

A wave is the movement of a ____1____ through a ____2____. Waves transfer ____3____ but not ____4____. A wave is caused by something that ____5____. **[5]**

5. Use some of the words below to complete the sentences.

direction frequency shape speed wavelength

At the boundary between two materials waves change ____1____. When this happens the ____2____ also changes (the ____3____ does not change) and this may cause a change in ____4____ of the wave. **[4]**

6. **(a)** Draw a diagram of a transverse wave.
(b) Mark on the wavelength.
(c) Mark on the amplitude. **[3]**

7. Some ocean waves have a wavelength of 200 m and 1 wave passes a rock every 10 seconds.
(a) What is the period of the waves?
(b) What is the frequency of the waves?
(c) Use the equation velocity = frequency x wavelength to work out the speed of the waves. **[3]**

Chapter

4 Electromagnetic radiation

The following topics are covered in this chapter:

- *Energy and intensity*
- *Wave communications*
- *Radio waves and microwaves*
- *Infrared and light*
- *Ultraviolet, X-rays and gamma rays*
- *The atmosphere and global warming*

4.1 Energy and intensity

The intensity of radiation

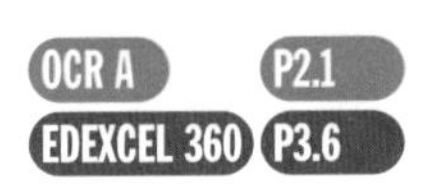

Electromagnetic radiation carries **energy**. When a beam of radiation strikes a surface, the **intensity** of the radiation is the energy arriving on a surface area of one square metre each second.

The intensity of electromagnetic radiation is the energy arriving at the surface each second on a surface area of one square metre.

The radiation is **emitted** from a **source** and travels towards a destination. On this **journey** the radiation **spreads out**, so the **further away** a **detector** is from the source, the less energy is detected. This is shown in Fig. 4.1. The **intensity** of the radiation can be increased by moving closer to the source.

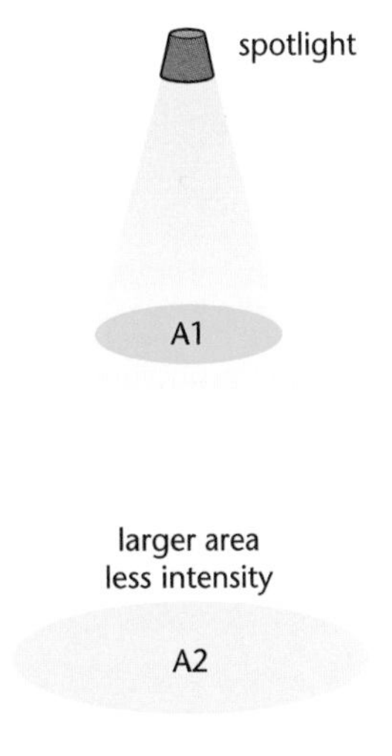

Fig. 4.1 Radiation spreads out from a source.

Area A2 is larger than A1, so the intensity is less. Intensity can also be increased by increasing the power (or total energy) of radiation from the source. One way of doing this is shown in Fig. 4.2.

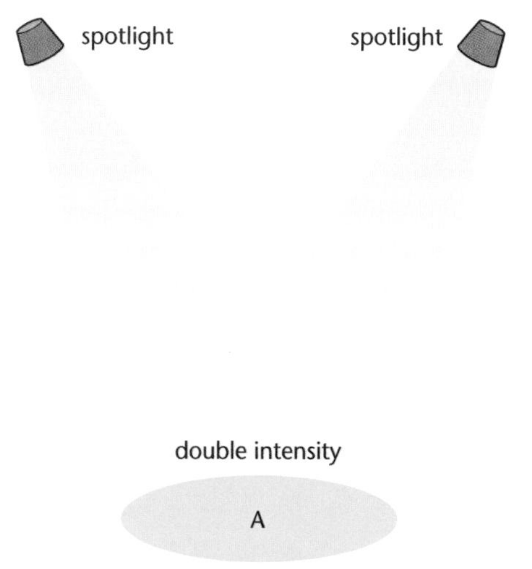

Fig. 4.2 The identical spotlights on the same area double the intensity.

Some materials **absorb** some types of electromagnetic radiation. The **further** the radiation travels through an absorbent material, the lower its **intensity** will be when it reaches the end of its journey. This is because more of the radiation energy is absorbed by the material. The material gains energy from the incident radiation and heats up.

On some journeys, electromagnetic radiation crosses a boundary between two different materials and is **reflected**.

The energy of radiation

The **spectrum** of electromagnetic radiation ranges continuously from **low-energy radio waves** to very **high-energy gamma rays**. The **electromagnetic radiation** comes in packets of **energy** called **photons**. The table lists the types of electromagnetic radiation in order of increasing photon energy.

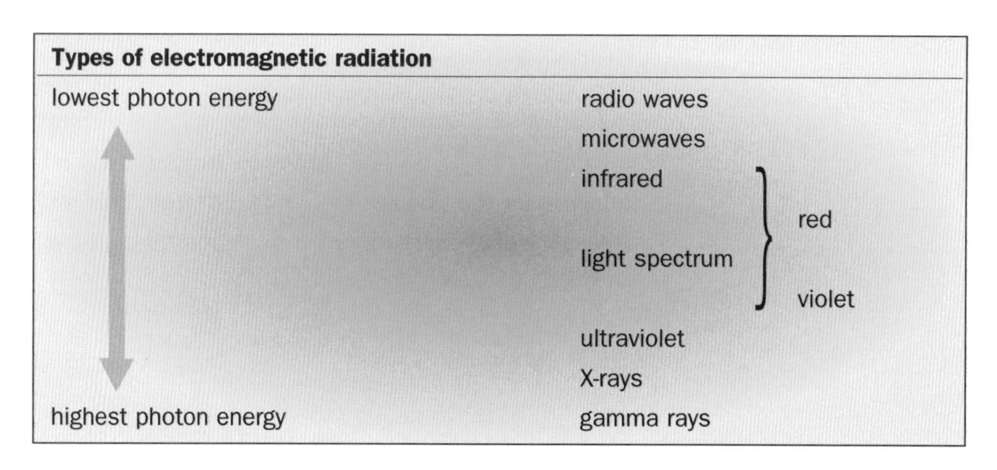

Types of electromagnetic radiation		
lowest photon energy	radio waves	
	microwaves	
	infrared	
	light spectrum	red
		violet
	ultraviolet	
	X-rays	
highest photon energy	gamma rays	

The boundary between each type of radiation is not a set value. For example, the highest-energy microwave photons merge into the lowest-energy infrared photons.

Electromagnetic radiation delivers energy in 'packets' called photons.

The **energy** arriving at a surface will depend on the **number of photons** striking the surface and the **energy of each photon**. For example, a photon of blue light has more energy than a photon of red light, so in Fig. 4.3, if the

Gamma ray photon energies and X-ray photon energies overlap. A gamma ray photon comes from the nucleus of an atom and an X-ray photon does not, but they can have the same energy.

number of red photons and the number of blue photons striking the area A is the same, then the blue spotlight illuminates the area A in the diagram with more energy than the red one.

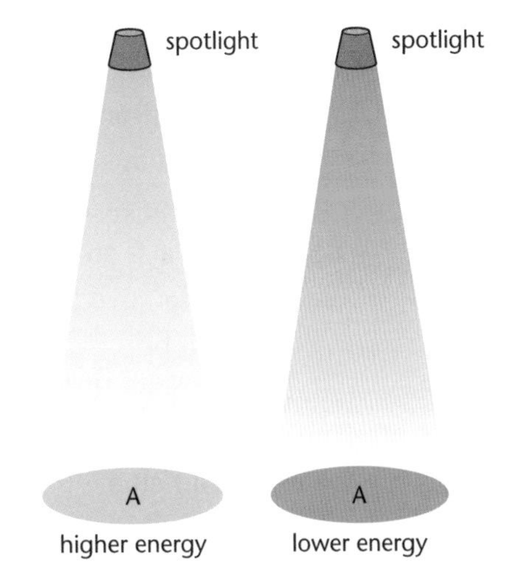

Fig. 4.3 Blue light has more energy than red light.

4.2 Wave Communications

Digital and analogue signals

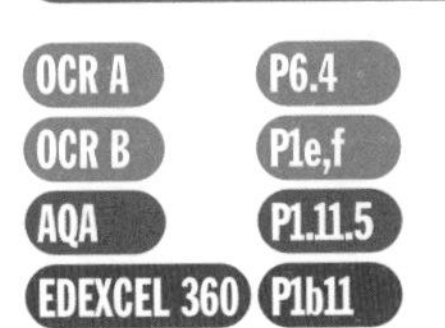

Electromagnetic radiation is used for **communications** and **transmission of information**. The waves that are used in this way are **radio waves**, **microwaves**, **infrared radiation** and **light**.

The idea of using a signal lamp to communicate was used in the 19th century. This method of long distance communication needed a code. One code used was **Morse code**, a series of long and short flashes of light for different letters of the alphabet. These signals can only be seen when visibility is good and for short distances.

Today we still use codes to send signals using electromagnetic radiation. There are two types of signal, **analogue** and **digital**. An analogue signal changes in frequency and amplitude all the time in a way that matches the changes in the voice or music being transmitted. A digital signal has just two values – which we can represent as **0 and 1**.

An analogue signal varies in frequency and amplitude. A digital signal has two values, 0 and 1 (or 'on' and 'off').

The signal (voice, music or data) is converted into a code using only the values 0 and 1. The signal becomes a stream of 0 and 1 values. These pulses are added to the electromagnetic wave and transmitted. The signal is received and then decoded to recover the original signal.

Both analogue and digital signals can pick up unwanted signals that distort the original signal. These **unwanted signals** are called **noise**. Digital signals can be cleaned up in a process known as **regeneration** because each pulse must be a 0 or a 1, so other values can be removed. Analogue signals can be **amplified**, but the noise is amplified too. This is why digital signals give a better-quality reception.

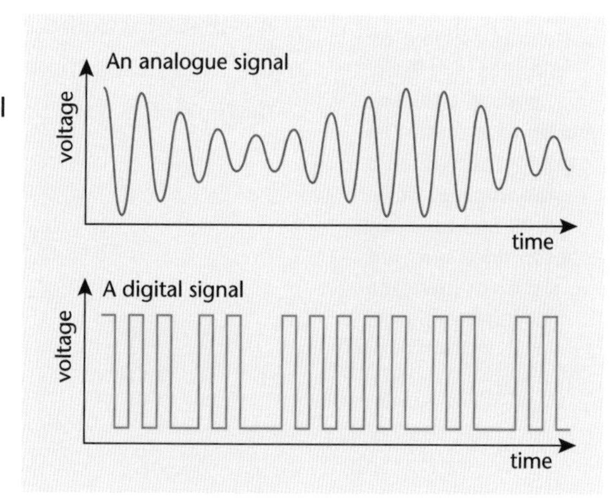

Fig. 4.4 How analogue and digital signals change with time.

Digital signals give a better-quality reception because noise on digital signals is more easily removed.

4.3 Radio waves and microwaves

Radio waves

A common mistake is to think that we can hear radio waves. We cannot hear any electromagnetic radiation. The radiation is used to carry a signal that is converted into a sound wave by the receiver.

Radio waves are the lowest-energy, lowest-frequency and longest-wavelength electromagnetic waves. They are produced when an **alternating current** flows in an **aerial** and they spread out and travel through the atmosphere. They are not strongly absorbed by the atmosphere. Another aerial is used as a detector and the waves produce an alternating current in it, with a frequency that matches that of the radio waves. Anyone with a receiver can tune it to this frequency to pick up the **radio waves** so they are suitable for **broadcasting** (for example, radio and TV programmes) to large numbers of people. An advantage is that this method of communicating does not require wires to transmit information. A disadvantage is that radio stations using similar transmission frequencies sometimes **interfere**.

Medium wavelength radio waves are **reflected** from the **ionosphere**, a layer of charged particles in the upper atmosphere, so they can be used for long distance communication.

Digital radio has better-quality reception as it uses digital signals and so does not have problems of noise and interference.

Microwaves

OCR A P2.1/2,P6.3/4
OCR B P1d,P1f,P5e
AQA P1.11.5
EDEXCEL 360 P1b11

For OCR A you need to know about the energy of electromagnetic radiation. For the other specifications you need to know about the energy, frequency and wavelength of the electromagnetic waves.

Microwaves are sometimes considered to be **very short radio waves** (high-frequency and high-energy radio waves).

Some important properties of microwaves are:

- They are **reflected** by **metal** surfaces.
- They **heat materials** if they can make **atoms** or **molecules** in the material **vibrate**. The amount of heating depends on the **intensity** of the microwave radiation, and the **time** that the material is exposed to the radiation.
- They pass through **glass and plastics**.
- They pass through the **atmosphere**.
- They pass through the **ionosphere** without being reflected.
- They are **absorbed by water molecules**, how well depends on the frequency (energy) of the microwaves.
- Transmission is affected by wave effects such as reflection, refraction, diffraction and interference.

Microwaves and water molecules

A **microwave** frequency (energy) can be selected which is strongly absorbed by **water molecules**, causing them to **vibrate**, and increasing their **kinetic energy**. This effect can be used to heat materials containing water, for example **food**. If the most strongly absorbed frequency (energy) is used in a **microwave oven** it only cooks the outside of the food because it is all absorbed before it penetrates the food. So the frequency (energy) used in a microwave oven is changed slightly to one that will **penetrate** about **1 cm** into the food. Conduction and convection processes then spread the heat through the food.

As our **bodies** contain water molecules in our **cells**, microwave oven radiation will heat up our cells and is very **dangerous at high intensity** because it will burn body tissue. The radiation is kept inside the oven by the **reflecting metal case** and **metal grid** in the door.

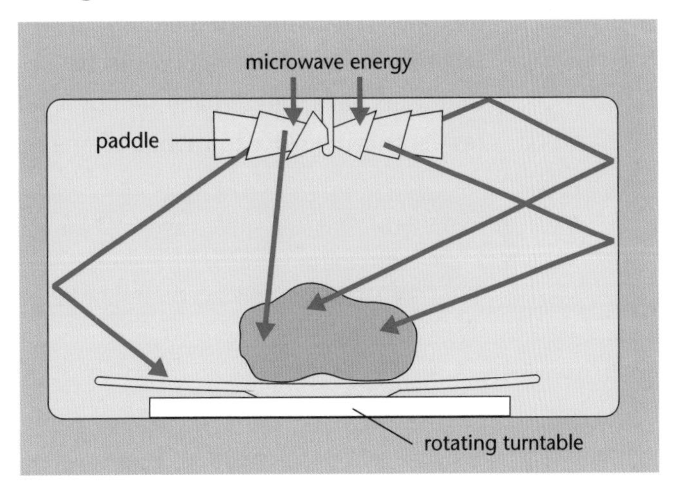

Fig. 4.5 A microwave oven.

Microwaves sent through the **atmosphere** will be absorbed by water so they can be used to **monitor rain**. The weaker the signal reaching the detector, the more rain the microwaves have passed through.

Microwave transmissions

Wireless technology uses **microwaves** and **radio waves** to transmit **information**. Advantages are:

- we can receive phone calls and email 24 hours a day
- no wiring is needed to connect laptops to the Internet, or for mobile phones or radio
- communication with wireless technology is portable and convenient.

Microwaves can be used to **transmit signals** over large distances if there are no obstacles between to reflect or absorb the beam. Another way to say this is that the transmitter and receiver are in **line of sight** (one can be seen from the other). This is why the transmitters are positioned high up, often on tall microwave masts. They cannot be spaced so far apart that, for example, hills or the curvature of the Earth stop the beam.

Microwaves are used to send signals to and from **satellites**. The satellites can relay signals around the Earth. Microwaves are used because they pass through the atmosphere and through the ionosphere. The signals may be for television programmes, telephone conversations, or monitoring the Earth (for example, weather forecasting).

There is more about diffraction in 3.5 Diffraction.

When microwaves are transmitted from a dish the wavelength must be small compared to the dish diameter to reduce diffraction – the spreading out of the beam. The dish is made of metal because metal reflects microwaves well.

Mobile phones use microwave signals. The signals from the transmitting phones reflect off metal surfaces and walls to communicate with the nearest transmitter mast. There is a network of transmitter masts to relay the signals on to the nearest mast to the receiving phone.

Mobile phones have not been in widespread use for many years, so there is not much data about the possible **dangers** of using them. The transmitter is held close to the user's head so the microwaves must have a small heating effect on the brain. There are questions about whether this could be dangerous, or whether it is not large enough to be a problem. So far studies have not found that users have suffered any serious ill effects. There may also be a risk to residents living close to mobile phone masts.

KEY POINT **Low-intensity microwave radiation, from mobile phone masts and handsets, may be a health risk, but there is disagreement about this.**

4.4 Infrared and light

Infrared radiation

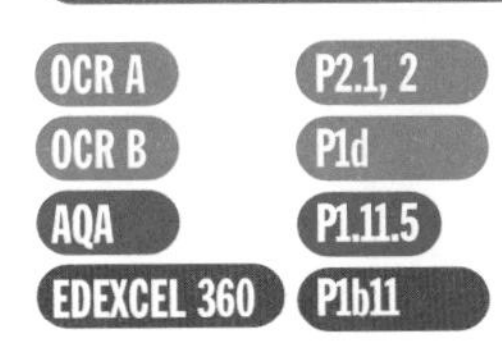

There is more about absorption and reflection in 3.5.

When **infrared** radiation strikes our **skin**, we feel **heat**. Infrared radiation is **absorbed** by **black** and **dull** surfaces and is **reflected** from **silver** and **shiny** surfaces. When it is absorbed, all the particles in the surface are heated. So infrared radiation is used for **cooking** the surface of food (the interior is then heated by convection and conduction). For the same reason we must be careful that intense infrared radiation does not burn our skin.

Infrared radiation is used in remote controls for televisions and other electronic appliances such as DVD and video recorders. Looking with a digital camera (which shows up infrared signals that our eyes cannot see) you can see the flashing infrared-emitting diode sending the signal. These signals cannot pass through solid objects but can reflect off walls and ceilings to operate the television.

Communications using infrared radiation and light

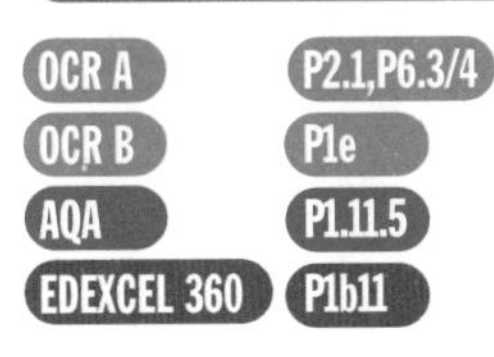

There is more about internal reflection in 3.5.

Infrared radiation and **light** travel along **glass optical fibres** by being **totally internally reflected**. The fibres are made with a core that refracts differently from that of the outer rim. The signal is reflected at the boundary as shown in Fig. 4.6.

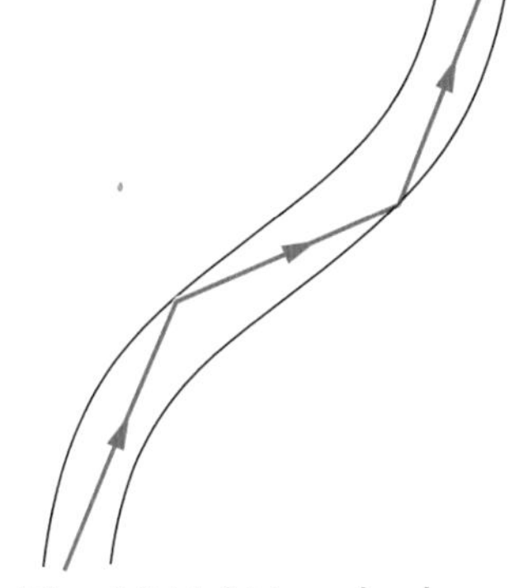

Fig. 4.6 Light travels along optical fibres.

Signals (**pulses of radiation**) can be sent for long distances using **optical fibres**. A stream of data can be transmitted very quickly. There is **less interference** than with microwaves passing through the atmosphere. It is also possible to use **multiplexing**, a way of sending many different signals down one fibre at the same time. Digital signals are used so that noise can be removed when the signal is regenerated.

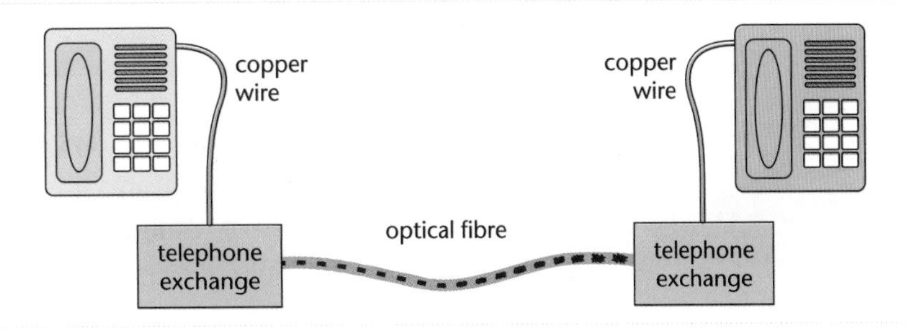

Fig. 4.7 Modern phones use optical fibres.

Lasers

OCR B P1g

Lasers produce a beam of radiation in which all the waves have the **same frequency** and are **in phase** (in step) with each other. Both infrared radiation and red light lasers are used as sources for fibre optic communications. They are also used in a **CD player.** Compact discs (CDs) store information digitally as a series of pits and bumps on its shiny surface. A laser beam is reflected differently from the pits and bumps and a detector is used to 'read' the different reflections as 0s and 1s – the information stored on the CD.

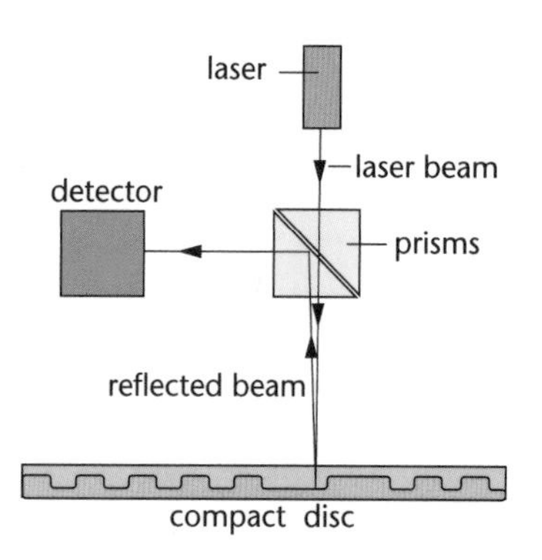

Fig. 4.8 A laser is used to play a CD.

Iris recognition

EDEXCEL 360 P1b11

Iris scanning is increasingly being used to **identify** people when high security is needed. The iris is so **individual** that the chance of mistaken identity is almost zero. A **digital camera** is used to take a clear high-contrast picture of the iris using both **infrared** radiation and **light**. A computer is used to locate the pupil (which appears very black in infrared illumination). The **patterns in the iris** are translated into a code. Two hundred points are compared in the iris (only 60 or 70 are used when matching fingerprints). The computer looks for features such as rings and freckles. The process takes about 5 seconds. The iris does not change much over a lifetime, and the technique works even when people wear contact lenses.

4.5 Ultraviolet, X-rays and gamma rays

Ionising radiation

OCR A P2.1/2
OCR B P1h
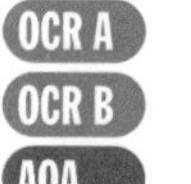
AQA P1.11.5, 11.6
EDEXCEL 360 P1b11

Ultraviolet radiation, X-rays and gamma rays are **high-energy** radiations that can **ionise** atoms they hit. If these atoms are ones inside the body the radiation can damage or kill cells. If the **DNA** in a **cell** is **damaged** it may **mutate**. This can cause cells to grow out of control – they become **cancer** cells.

Ultraviolet radiation

OCR A P2.1/2
OCR B P1h
AQA P1.11.5
EDEXCEL 360 P1b11

Ultraviolet radiation from the Sun travels towards the Earth. The **highest-energy** (highest-frequency) ultraviolet radiation is stopped by the **ozone layer**, a layer of the gas ozone in the upper atmosphere.

Unfortunately, using **CFC gases** in aerosol cans (as a propellant to force the contents of the can out of the nozzle) and as a refrigerant in refrigerators and freezers, caused CFC gas to increase in the atmosphere. This pollution **reacted** with the **ozone** and reduced the amount in the ozone layer.

Above the Antarctic an '**ozone hole**' developed and began to grow. International agreements have stopped the use of CFCs but the ozone layer will take time to recover. Where the ozone layer has been depleted, living organisms, especially animals, suffer more harmful effects from ultraviolet radiation.

The **lower-energy** ultraviolet radiation that passes through the ozone layer can cause **skin cancer** and damage the lens of the eye. The number of cases of skin cancer rose when people started to spend more time in the sun. **Light skins** are more **easily damaged** than **dark skins**. Dark skins absorb more of the ultraviolet radiation close to the surface; light skins allow the radiation to penetrate further into the body tissues.

Campaigns warn people to stay out of the sun during the hottest part of the day, and to cover up with a hat and clothes. They should also use a **high-protection factor sunscreen**, which reduces the ultraviolet radiation reaching the skin.

Sunscreens have a **sun protection factor (SPF)** number, which relates to how long a user can stay out in the sun. To find out the time you can stay in the sun wearing the sunscreen, **multiply the time** you could safely stay in the Sun **without** the sunscreen by the SPF. For example, if a person could safely stay in the sun for 10 minutes, using a sunscreen with SPF 15 means they can can safely stay in the sun for 150 minutes. It is important:

- not to miss any area of skin
- to put the cream on in a thick enough layer
- to reapply it according to the instructions.

Ulltraviolet blocking **sunglasses** are recommended to protect the **lens** of the **eye** from damage.

Ionising radiation can cause cells to turn cancerous. Ultraviolet radiation from the Sun can cause skin cancer.

Using ultraviolet radiation

Ultraviolet radiation is used to detect **forged bank notes**. Bank notes have some features, for example the number of pounds, that **fluoresce** in ultraviolet radiation. This means they absorb the ultraviolet and re-emit visible light. You may have seen a detector in shops. The banknotes are placed under the light and the number shows up brightly, because it fluoresces if the bank note is genuine.

X-rays

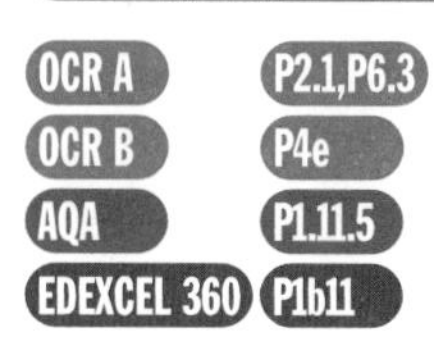

X-rays have **high energy** and can pass through the **body tissues**. They are **stopped** by **denser** materials such as the **bones** and by pieces of metal. They can be used to scan the body and show breaks in the bones. The diagram shows how a photographic plate can be placed behind the patient and X-rays directed towards the patient. The plate will darken where the X-rays strike it and leave white shadows of the bones, where the X-rays are absorbed.

The **operator** stands behind a **lead screen** so that he or she is not exposed to X-rays each time a patient is X-rayed. The patient should not receive too many X-rays. They are used when the **benefit** (for example, finding a broken bone) is greater than the **risk.**

In the 1950s, all pregnant women were X-rayed to check on the development and position of the baby. This caused a few cancers among children. The benefits were not greater than the risks and routine X-rays were stopped.

X-rays are used for **security scans** of passengers' luggage. They pass through the suitcase and clothes, but metal items and batteries stop the X-rays and show up as shadows on the screen.

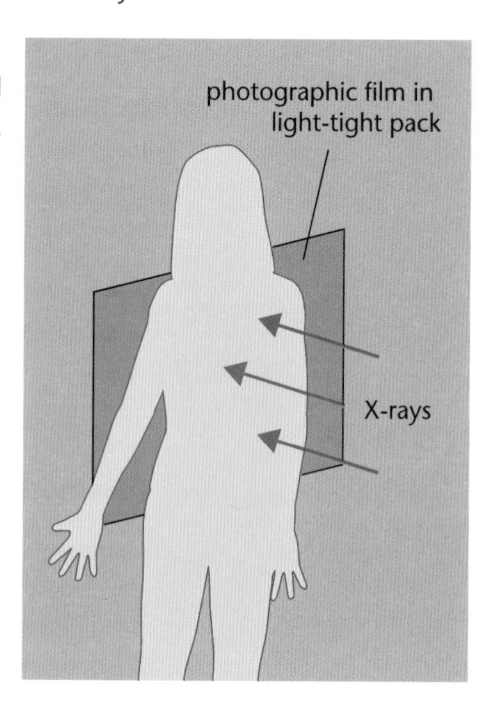

Fig. 4.9 Taking an X-ray picture

4.6 The atmosphere and global warming

The carbon cycle

Plants absorb **carbon dioxide** from the air and water around them. Together with **water** and energy (**sunlight**,) plants make the sugar called **glucose** in their leaves, and release **oxygen**.

This process is called **photosynthesis**.

Photosynthesis takes place in the leaves of green plants:

carbon dioxide + water + sunlight → glucose + oxygen

Plants make other carbon compounds from the glucose. When animals eat plants and then other animals eat the animals, the carbon is transferred to the animals.

Plants and animals give out carbon dioxide when they respire. **Respiration** is the reverse of photosynthesis, the living cells of plants and animals release the **energy** stored in **glucose**. To do this they use **oxygen** from the atmosphere and give out **water** and **carbon dioxide**.

When living things die **decomposers**, for example, soil bacteria and fungi return carbon dioxide to the air.

These reactions are part of the **carbon cycle**, see Fig. 4.1, which describes how carbon is transferred from and to the atmosphere. Hundreds of millions of years ago there was more carbon dioxide in the atmosphere, but then green plants started to photosynthesise and release oxygen, making life possible for animals. **Fossil fuels**, and rocks like **limestone** and **chalk**, contain a lot of carbon that was once part of living things, and before that, was part of carbon dioxide in the atmosphere.

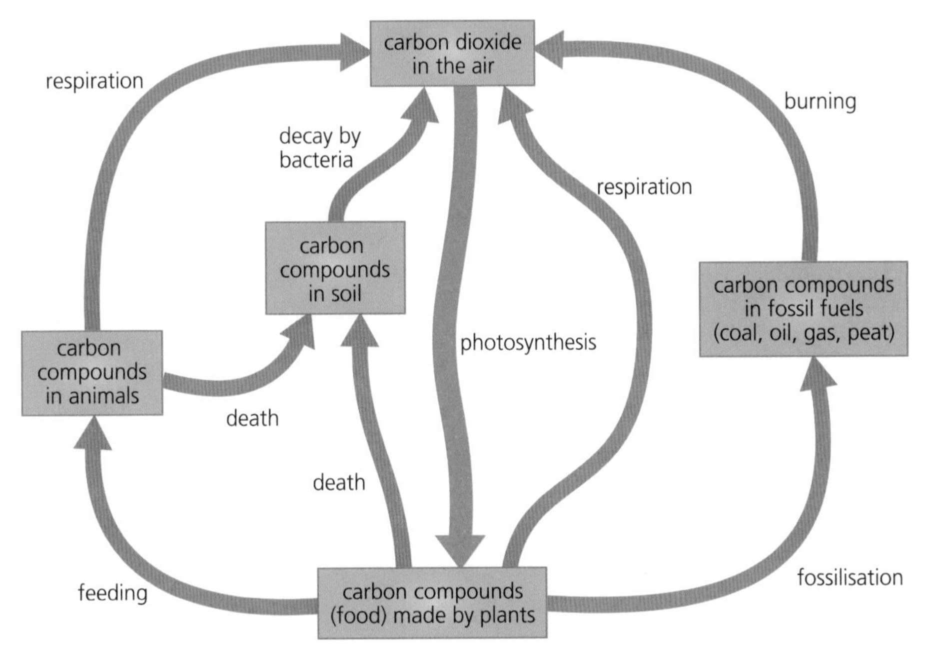

Fig. 4.10 The carbon cycle.

Gases in the atmosphere

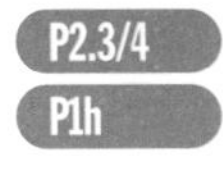

P2.3/4
P1h

The Earth is surrounded by an atmosphere made up of different gases. The radiation from the Sun that can pass through the atmosphere is infrared radiation, light and ultraviolet radiation. Infrared and ultraviolet radiation are close to light in energy and frequency.

The radiation from the Sun provides energy for photosynthesis and warms the Earth's surface.

The Earth emits infrared radiation, but this radiation is at a low energy (low frequency). These energies are absorbed by some gases in the atmosphere such as **carbon dioxide**, **water vapour** and **methane**. This keeps the Earth warm. It is called the **greenhouse effect**. Without the greenhouse effect the Earth would be much colder – probably too cold for some species to survive.

In the last 200 years, the amount of carbon dioxide in the atmosphere has steadily increased. This is partly because we burn so much **fossil fuel**. We have also cleared many forests so that fewer trees are using the carbon dioxide for photosynthesis. This means that more of the infrared radiation is being absorbed and the Earth is warming up. This **global warming** causes **climate change**, which, scientists agree, is already taking place. There is less agreement about how much change is likely and what effect it will have. It could result in:

- extreme weather conditions in some regions
- rising sea levels due to melting ice and expansion of water in oceans, which may flood low-lying land
- some regions no longer able to grow food crops.

High-energy (high-frequency) infrared radiation from the Sun can pass through the glass into a greenhouse. Low-energy (low-frequency) infrared radiation from the plants cannot pass out through the glass. This keeps the greenhouse warm – the greenhouse effect.

Dust in the atmosphere

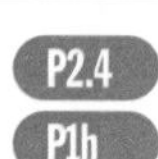

When a volcano erupts it produces a lot of gases and dust, which spread around the atmosphere. These reflect the radiation from the Sun and cause the Earth to cool.

If factories in cities produce large amounts of smoke and dust, these can reflect the radiation emitted from the city and keep it warmer.

HOW SCIENCE WORKS

OCR A P2.5
OCR B P1d
AQA P1.11.5
EDEXCEL 360 P1b.11

Mobile phones

Before 1985, no one in the UK had a mobile phone, but by 2005, as shown in the table, 50 million people in the UK had, or regularly used, a mobile phone.

Year	Mobile phone users in the UK
1985	0
1990	Fewer than 1 000 000
1995	4 500 000
2000	25 000 000
2005	50 000 000

This means that scientists do not have enough data to say if using a mobile phone is harmful. They are studying sample populations to see if there are harmful effects, but it will be some time before they have conducted enough studies to be sure.

The **precautionary principle** says that we if we don't know about the effects of something we should not take the risk – 'Better safe than sorry.' In 2000, a group set up by the UK government called 'The Independent Expert Group on Mobile Phones' reported their findings on mobile phone safety. The following text summarises a Department of Health leaflet.

Risks

The radio waves that are received and sent by mobile phones transmit in all directions, to find the nearest base station. This means that some of the radio waves will be directed at the head of the person using the phone. These waves are absorbed into the body tissue as energy, which can eventually cause a very small rise in temperature in the head.

This effect is measured using specific absorption rates (SARs), which is a measure of the amount of energy absorbed by the body. The higher the SAR, the more energy your body is absorbing and the higher the rise in temperature.

Present research shows that the radio waves from mobile phones are sufficient to cause a rise in temperature of up to 0.1 °C. This does not pose a known risk to health. Some mobile phones have better SARs than others; you can find this information from your mobile phone manufacturer or retailer.

Children are thought to be at higher risk of health implications from the use of mobile phones. This is because their skulls and cells are still growing and tend to absorb radiation more easily.

Recommendations
You can minimise your exposure to radio waves:

- Only make short calls on your mobile phone.
- Children should use mobile phones only if absolutely necessary. Find out the relative SARs before you buy a new mobile phone.
- Keep your mobile phone away from your body when it is in standby mode.
- Only use your phone when the reception is strong. Weak reception causes the phone to use more energy to communicate with the base station.
- Use a mobile phone that has an external antenna. This keeps the radio waves as far away from your head as possible.

HOW SCIENCE WORKS

Sun and skin cancer

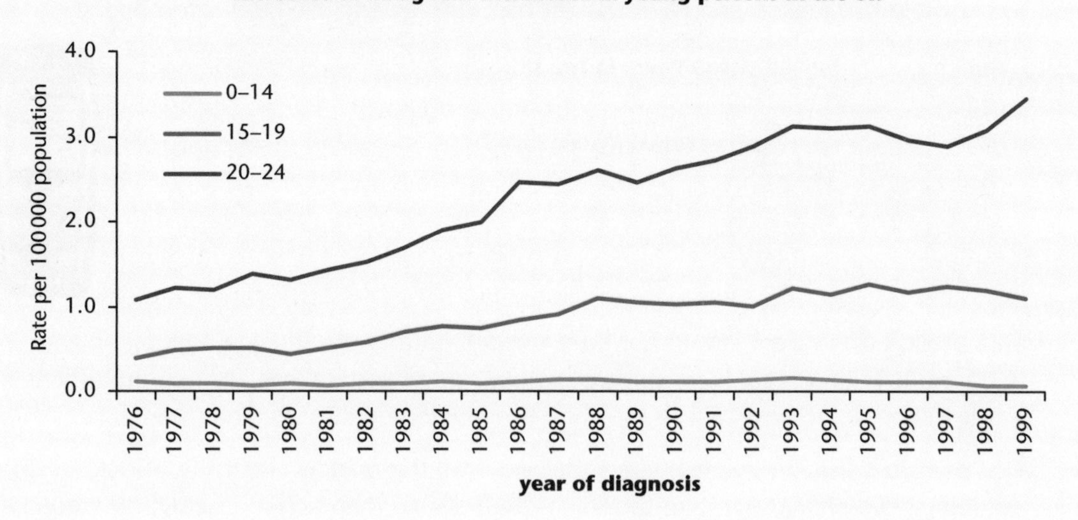

This graph shows that the number of cases of a type of skin cancer – malignant melanoma – has been steadily increasing. The graph may give the impression that 20–24-year-olds are at the most risk from the Sun. In fact this is because the damage to a child's skin from the Sun can cause cancer when they are older. In the UK we can learn from what has happened in Australia.

Australia is the country with the highest incidence of skin cancer in the world. Two out of three Australians will be treated for some form of skin cancer during their lifetime. This is partly because Australia has a generally light-skinned population with an outdoor lifestyle, and also it has clear skies, a depleted ozone layer and is close to the equator.

Data shows cases of melanoma began to rise in the 1930s. In the early 1980s, because the number of deaths from skin cancer was rapidly rising, the 'Slip! Slop! Slap!' programme encouraged fair-skinned Australians to protect themselves by slipping on a shirt, slopping on sunscreen and slapping on a hat. In the 1990s, Sun-protective clothing became popular because it was much easier to use than sunscreen. There was a 50% reduction in people getting sunburnt between 1988 and 1998.

In Australians under 60 years of age, the programme began to show success, when cases of melanoma fell between the mid-1980s to the mid-1990s. (Cases continued to increase in older Australians who had been exposed to the Sun before the 1980s.) Scientists hope that once the children born in the 1980s reach the age of 60, melanoma cases will have fallen to a very low level again. At the moment, after more than 60 years of steadily increasing deaths from melanoma, the trend has finally changed. This suggests younger Australians are now better at protecting themselves from the Sun.

1. Scientists are setting up a study to see if brain tumours in children are increased by using mobile phones.

Here are 3 possible samples for a 10 year study:

A 100 children aged between 8 and 12. 50 use a mobile phone 50 do not.

B 1000 children aged between 0 and 15. 500 use a mobile phone at the start of the trial 950 use one by the end.

C 1000 children aged between 8 and 12. 400 use a mobile phone at the start, 600 use one at the end of the trial.

(a) Give an advantage and disadvantage of sample A. **[2]**

(b) Give an advantage and disadvantage of sample B. **[2]**

(c) Explain how sample C could be used to produce a better sample of 800 for comparison. **[2]**

(d) Explain whether you think 10 years is a reasonable time for the study. **[1]**

(e) Some children use a mobile phone for several hours of conversation a day, others only use it to text. Explain how this will affect the trial. **[2]**

2. From the graph of incidence of melanoma, on average, how many 20–24 year olds out of 100 000 had malignant melanoma in **(a)** 1985 **(b)** 1995? **[2]**

Exam practice questions

1. Which of the following is ionising radiation?
   - **A** infrared radiation
   - **B** light
   - **C** microwave radiation
   - **D** ultraviolet radiation **[1]**

2. Which radiation is best suited to communication between Earth-based stations and orbiting satellites?
   - **A** light
   - **B** microwave
   - **C** radio
   - **D** ultraviolet **[1]**

3. How does light travel along optical fibres?
   - **A** by diffraction
   - **B** by dispersion
   - **C** by refraction
   - **D** by total internal reflection **[1]**

4. Use the words from this list to complete the sentences below. (Use words, once, more than once, or not at all.)

   analogue **digital** **infrared** **light** **noise** **optical fibre** **regenerated**

   A signal can be transmitted using ___1___ radiation through an ___2___. When a signal is amplified any ___3___ it has picked up is also amplified. This is a problem with ___4___ signals. A signal that is a code of 0s and 1s is called a ___5___ signal. Small added ___6___ signals do not have the value 0 or 1 and can be set to 0 giving a cleaned up signal when the signal is ___7___. **[4]**

5. This is a list of statements about electromagnetic radiation. Write **T** for the **true** statements and **F** for the **false** statements.
   - **(a)** Gamma rays have higher energy photons than microwaves.
   - **(b)** Ultra violet radiation is ionising.
   - **(c)** The intensity of the radiation does not depend on the energy of the photons.
   - **(d)** Microwaves are reflected by glass.
   - **(e)** Infrared radiation and light travel along glass fibres by being refracted.
   - **(f)** X-rays pass through soft tissues but are absorbed by bone. **[3]**

Exam practice questions

6. Use the words from this list to complete the sentences below. (Use words once, more than once or not at all.)

absorbed grid metal microwaves reflect
reflects transparent vibrate water

In a microwave oven, molecules in the food are made to ___1___ by the ___2___. This makes them heat up. The oven casing is made of ___3___ which ___4___ the ___5___. The door has a metal ___6___ in it to ___7___ the microwaves. Plastic and glass are used to contain the food because they are ___8___ to microwaves.

7. Join the graphs of these signals to their description.

Graph of signal

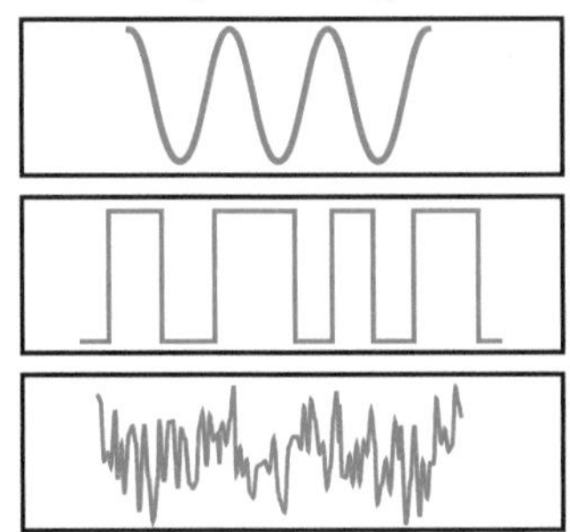

Description of signal

noisy

analogue

digital

[2]

8. **(a)** Explain what is meant by 'wireless technology'

(b) What are **two** advantages of wireless technology? **[3]**

Chapter 5 The Earth and beyond

The following topics are covered in this chapter:

- *The Earth*
- *The Solar System*
- *Gravity and orbits*
- *Beyond the Solar System*
- *Observing and exploring*
- *The expanding Universe*

5.1 The Earth

The structure of the Earth

OCR A P1.1/2/4

The **oldest** rocks on the **Earth** are about **4 thousand million years old**, which tells us that the Earth must be older than this. The rocks provide evidence for changes in the Earth.

The rock cycle

Rocks are **eroded** and the pieces are washed down into lakes and seas to form layers of **sediment**. These layers may contain dead animals and plants. The sediments are compressed into **sedimentary rocks** containing **fossils**. The newest rock is the top layer. Eventually all the continents would be worn down to sea level, but mountains are being continually formed. Over time, the rock layers may be lifted up and **folded** into **mountains.** They are then eroded again. Rocks can be dated by the order of the **layers**, the **fossils** they contain and by **radioactive dating**. (Because radioactive isotopes decay over time, the smaller the amount left in a rock, the older the rock.)

It was once thought that the mountains and valleys on the Earth's surface were formed by the shrinking of the Earth's crust as the newly formed hot Earth cooled down. This theory has been replaced by the **theory of plate tectonics**.

Plate tectonics

The Earth is made up of a **core**, a **mantle** and a **crust** as shown in the diagram.

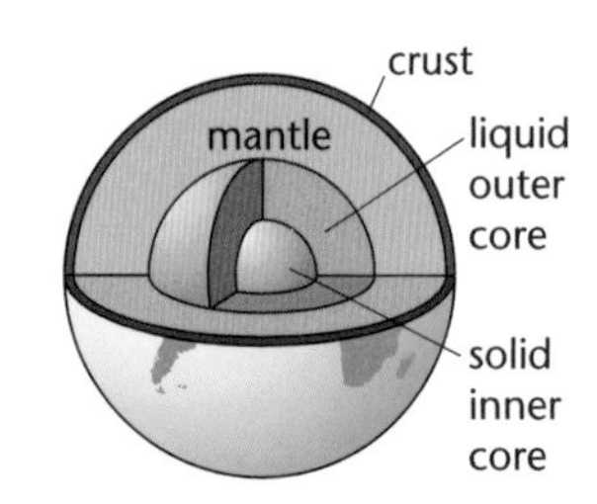

Fig. 5.1 The Earth's structure.

Once the world was mapped, people looked at the shape of the continents and thought they looked like pieces of a jigsaw puzzle. Alfred Wegener took this further, saying that fossils of plants and animals in America and Africa were similar, but modern plants and animals were not, suggesting that the continents had been joined in the past. He had no theory to account for how the continents moved or evidence that they were actually moving so the theory was not believed.

The theory of continental drift was first suggested by Alfred Wegener in 1915, but, owing to lack of evidence, it was not until 1967 that the theory of plate tectonics was accepted.

Wegener was also not believed because he was an outsider, not one of the community of geologists.

The theory says that the Earth's crust and the upper part of the mantle are made of a number of large plates, called **tectonic plates**. The plates are on top of the mantle because they are less dense than the mantle. There are **oceanic plates** under oceans and **continental plates** forming continents. Wegener was not believed because he had no explanation for why the continents move. The plate tectonic theory explains that:

- The **mantle** behaves like a very **thick liquid**.
- Heat from **radioactive decay** causes very slow-moving **convection currents** in the **mantle**.
- As the material from the mantle rises, it **melts** and becomes liquid **magma.**
- Magma flows out of the mid-ocean ridges, forming new rock.
- The **sea floor** spreads by about **10 cm a year** and this causes the continents to move apart.
- At other places where two **plates collide**, rocks are pushed up, forming new **mountain ranges**.
- The plates are always moving. At the plate boundaries earthquakes and volcanic eruptions occur. In some places mountains are formed, in other places oceanic trenches form.

This diagram shows the direction of movement of the plates.

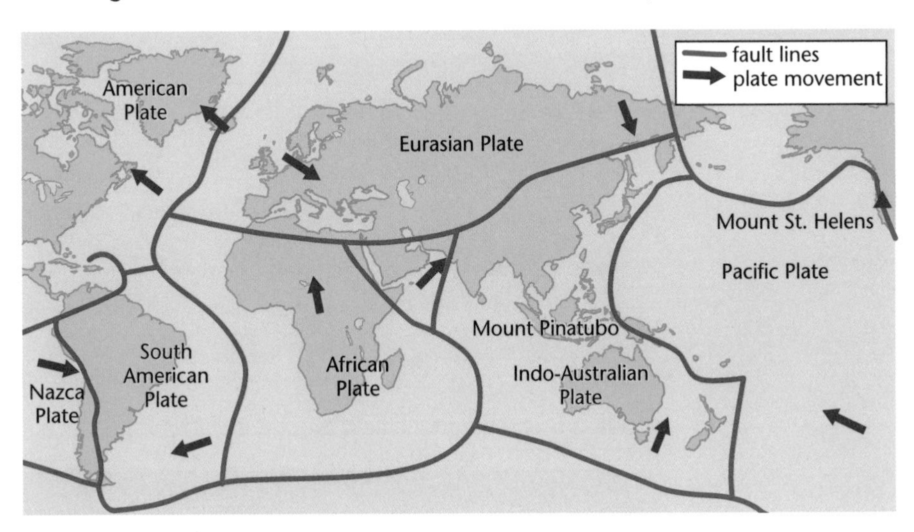

Fig. 5.2 The tectonic plates.

Evidence for this is found in the new rocks at the **ocean ridge**. The new rocks are rich in **iron**. Every few thousand years the Earth's **magnetic field** direction **reverses**. As the rocks solidify they are magnetised in the direction of the Earth's magnetic field. So the rocks contain a **magnetic record** of the Earth's field.

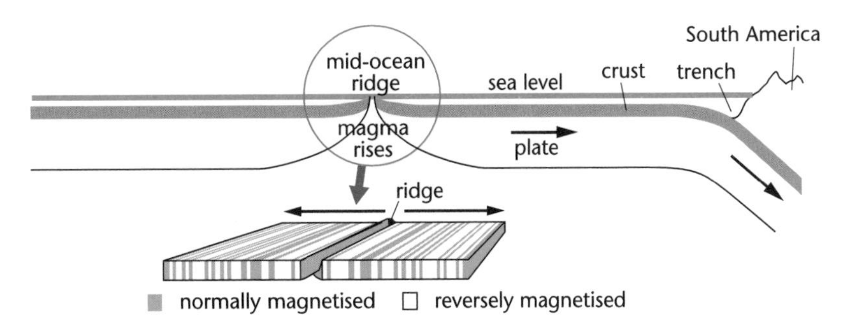

Fig. 5.3 The sea floor contains a magnetic record.

The Earth's magnetic field

OCR B P2e

The north pole of a magnet is attracted to the south pole of another magnet. The name 'north pole' of a magnet means 'North of the Earth-finding end' of a magnet, in other words the North Pole of the Earth behaves as the south pole of a bar magnet.

The **Earth** has a **magnetic field** as shown in the diagram. It can be thought of as if a large imaginary bar magnet lies inside the Earth with the south pole of the bar magnet at the Earth's magnetic North.

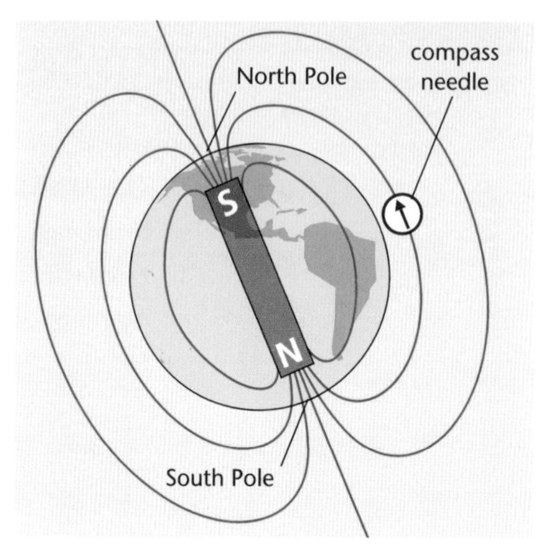

Fig. 5.4 The magnetic field of the Earth.

An **electric current** in a **coil of wire** generates a **magnetic field** as shown in Fig. 5.5. **Plotting compasses** can be used to investigate the **direction** of a **magnetic field**. The compass needle is a small magnet that will line up with the magnetic field.

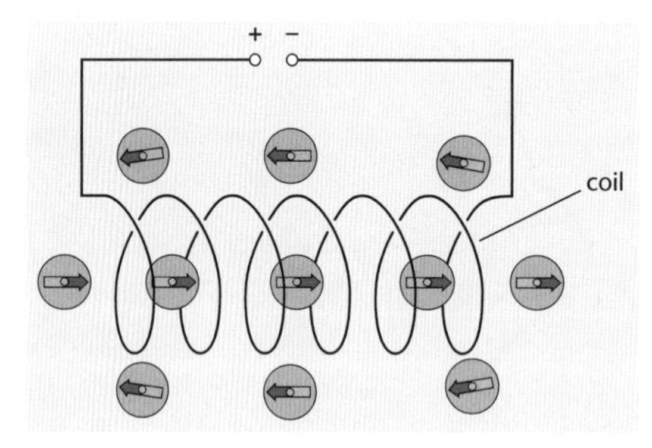

Fig 5.5 The magnetic field around a coil when a current passes.

The core of the Earth contains a lot of molten iron. As the Earth spins, electric currents flow in circles in the Earth's core and generate the magnetic field.

Cosmic rays

OCR B P2e

Cosmic rays are **ionising radiations** from space. They are fast-moving, charged particles. When they hit the atmosphere they can create **gamma rays**.

Moving charged particles (protons, electrons and ions) change direction when they move through a magnetic field.

Cosmic rays change direction when they reach the **Earth's magnetic field**. They **spiral** round the **Earth's magnetic field lines**, and come down to Earth at the Poles. As they pass through the atmosphere they collide with gas molecules. As a result, the gas molecules emit coloured light. At the North Pole the lights are called the **aurora borealis**.

Solar flares

A solar flare is a violent eruption in the Sun's atmosphere. Clouds of charged particles from the Sun are thrown out at high speed. They produce strong, changing magnetic fields. Some of these clouds of particles can travel towards the Earth.

The fastest cloud of charged particles measured arrived at the Earth only 15 minutes after leaving the Sun. The **Earth's magnetic field** deflects many of the particles and **protects** us on the **Earth**, but the operation of artificial satellites in orbit around the Earth can be affected. If the **electronic circuits** in **artificial satellites** are **damaged**, satellites can break down, affecting telecommunications, navigation and weather prediction.

Astronauts would have very little time to shield themselves from the lethal radiation from this ionising radiation – and a spacecraft offers little protection.

The magnetic field of the cloud of particles **disturbs** the **magnetic field** of the **Earth**, so that there are changing magnetic fields on the Earth. In the past, **electricity distribution networks** have **broken down** because of surges of electric current caused by the changing magnetic fields.

We now have satellites to give us some advance warning of clouds of particles heading towards Earth, and some equipment can be safely shut down.

The Moon

The current theory of how our **Moon** was formed is that the **Earth collided** with another **planet**, about the size of Mars. Most of the heavier material of the other planet fell to Earth after the collision and the iron core of the Earth and the other planet merged. Some less dense material was thrown into orbit and formed the Moon. We have the following evidence for this:

- Samples of Moon rocks have been brought back to Earth by astronauts and the Moon **rocks** have the **same composition** of isotopes as Earth rocks – unlike rocks from other planets and moons.
- The Moon is completely made of **less dense** rocks – there is no iron core – unlike other planets, moons and asteroids.
- The Moon has **no recent volcanic activity** but its rocks are **igneous**.

5.2 The Solar System

The planets and other objects

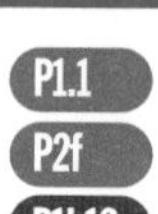

For OCR B you need to know the planets in order from the Sun: Mercury, Venus, Earth, Mars, Jupiter, Saturn, Uranus and Neptune and Pluto.

Pluto has been re-classified as a dwarf planet.

The **Solar System** was formed over a very long time from **clouds** of **gases** and **dust** in space, about **five thousand million years ago**.

The **planets** orbit around the **Sun** (the star at the centre of the Solar System). Some of the planets are orbited by one or more **moons**. An object that orbits another is called a **satellite**. The Moon is a natural satellite of the Earth – and we have placed a number of **artificial satellites** in orbit around the Earth, and some around the Sun and other planets.

Asteroids are rocks that orbit the Sun. They vary in size – the largest is almost 1000 km across, some are 100 km across but many are as small as pebbles. They have been around since the formation of the Solar System. Most of these are between Mars and Jupiter. Jupiter is the largest planet and there is a large gravitational force towards it. This has prevented the formation of a planet between Mars and Jupiter, in the asteroid belt.

There are many **comets**. Some take less than a hundred years to orbit the Sun, others take millions of years. They are made of ice and dust. Most have a nucleus of less than about 10 km, which vaporises and becomes a cloud thousands of miles across when the comet is close to the Sun. Comets spend most of their time far from the Sun – much further away than the planet Pluto. **Meteors**, or shooting stars, are caused by dust and small rocks, usually from a **comet**. When the Earth passes through this debris in its orbit around the Sun, the dust and rocks are attracted by **gravity** towards the Earth. As they pass through the atmosphere, the pieces are heated and glow. Any pieces that land on the Earth are called **meteorites.**

Near Earth objects

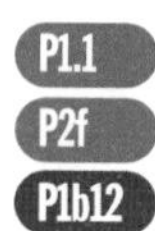

Near Earth objects (NEOs) are **comets** and **asteroids** that have been affected by the gravity of other planets so that they are now in an orbit that brings them close to the Earth. They are studied because their chemical composition is different from that of the Earth and they can give us information about the formation of the Solar System. Some of the NEOs could one day **collide** with **Earth**. The **craters** on the **Moon** are evidence of collisions in the past. **Craters** on the **Earth** have been mostly been **eroded** away, but the Barringer meteor crater can clearly be seen in the Arizona desert in the USA. A **collision** with a **large NEO** would result in a **crater** being formed and hot rocks being thrown up into the atmosphere. There would be widespread fires and the sunlight would be blocked by dust. This would cause **climate change** and many species would become **extinct**. The dust, containing unusual elements from the NEO, would settle as a layer onto the Earth's surface. The layer would be included in the new sedimentary rocks being formed, providing a record of the collision. There would be a smaller number of types of fossils above the layer (after the collision) than below the layer (before the collision). Evidence such as this points to the extinction of the dinosaurs being caused by a collision with a NEO.

There is evidence that the extinction of the dinosaurs was caused by the collision of a large NEO with the Earth.

Monitoring NEOs

Surveys by **telescope** observe and record the paths (trajectories) of all NEOs. They can be monitored from Earth or by satellite. We can make sure that we have advance warning of a collision. The idea of deflecting an NEO using explosions is being considered. At present, scientists are collecting information about the composition and structure of NEOs. The possibility of destroying one is only at the planning stage.

5.3 Gravity and orbits

Gravity

P2f/2g/5a
P3.13.3
P1b12

The force of gravity is a force of attraction between two objects with mass. The force gets larger if the mass of either of the objects is increased. The force gets smaller as the distance between the two masses is increased.

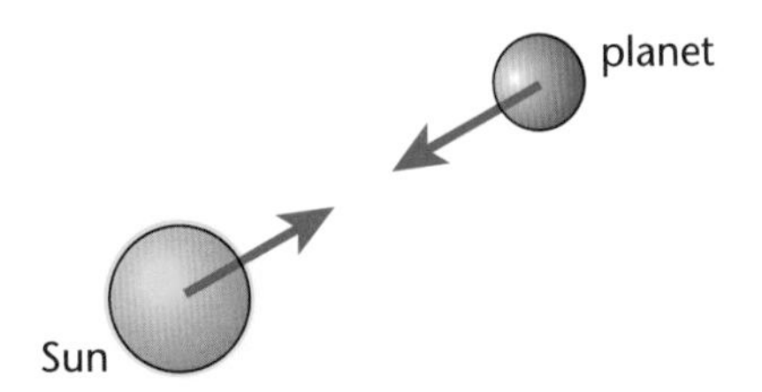

Fig. 5.6 The Sun attracts the planet with the same force as the planet attracts the Sun.

The **Sun** attracts the Earth and all the other **planets**. The **force of gravity** on a planet from the Sun keeps it moving in **orbit** around the **Sun**.

For an object to move in a circle, there must be a force on it towards the centre of the circle. This force is called the centripetal force.

If an object moves closer to the Sun, the gravitational force on it will increase and it will speed up.

Comets have very **elliptical orbits**. The force that keeps them in orbit is the gravitational force of attraction to the Sun but their distance from the Sun changes as shown in Figure 5.7. The **force** on the **comet** is **largest close** to the **Sun**, where the distance is smallest. The **speed** of the **comet** is much **larger** when it is **closer** to the **Sun**.

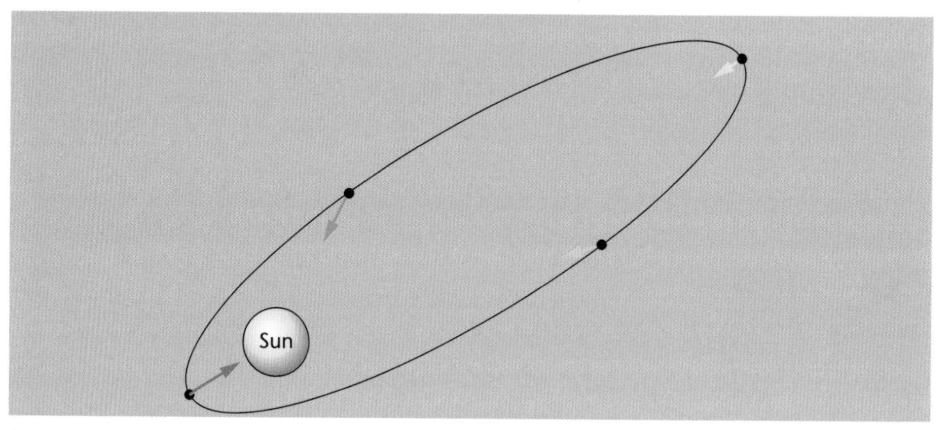

Fig. 5.7 A comet's orbit.

> **KEY POINT**
>
> **Weight is the gravitational force of attraction on a mass on the Earth. This can be written as the equation:**
>
> **Weight = mass × gravitational field strength**
> **The units of gravitational field strength are N/kg.**
> **Or**
> **Weight = mass × acceleration of free-fall**
> **The units of acceleration are m/s^2.**

On other planets, the gravitational field strength will be different – so **weight changes** but **mass** stays **the same**. In deep space where there is no gravity, objects will have zero weight.

5.4 Beyond the Solar System

Stars

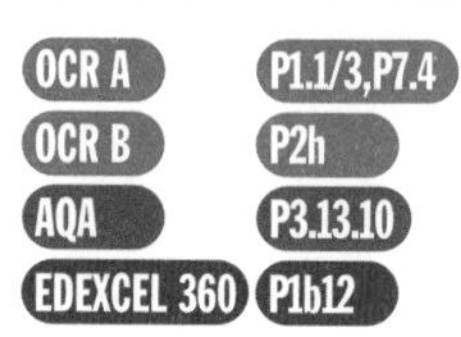

Stars begin as large clouds of **dust**, **hydrogen** and **helium** – an **interstellar gas cloud**.

- The cloud is sometimes called a **nebula**. **Gravitational forces** between the particles make the nebula **contract**. This makes it heat up and it is now a **protostar**. As the core gets hotter, the atoms collide at high speed, losing their electrons.
- When the temperature is high enough, the **hydrogen nuclei fuse** together to form **helium nuclei**. This process is called **thermonuclear fusion**. When light nuclei fuse they release energy. **Light** and other **electromagnetic radiation** is also released. A **star** has been formed.

- The **star** is one of a large number of **main sequence** stars. The high pressure in the core is balanced by the attractive gravitational forces. Our **Sun** is a main sequence star. Stars spend a long time fusing hydrogen. Our Sun will do this for ten thousand million years. Eventually a star converts most of its hydrogen to helium. What happens next depends on the mass of the star.
- A **star** of **mass similar** to that of our **Sun** will eventually cool, becoming redder and expand to form a **red giant**. The core will contract and **helium** will fuse to form **carbon** and **oxygen**. After all the helium has fused, the star will contract and the outer layers will be gently lost. As these outer layers move away they look to us like a disc, which we call a **planetary nebula**. The remaining core becomes a small, dense, very hot **white dwarf**. This remnant core will then cool over an incredibly long time and eventually it will become a **black dwarf**.

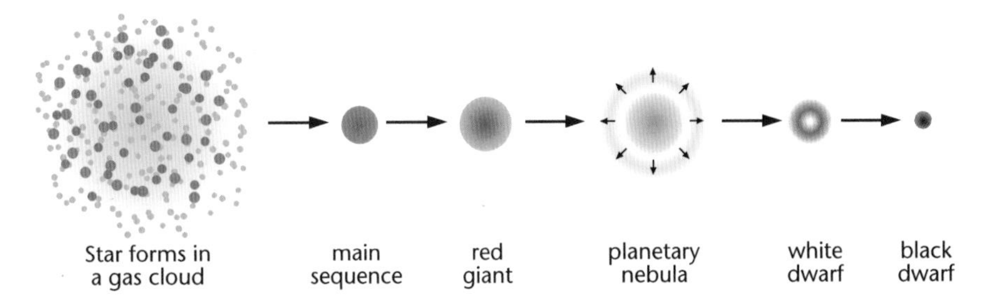

Fig. 5.8 The life cycle of a star of mass similar to that of our Sun.

All the elements heavier than helium were created in stars and supernovae. Everything on Earth – including us – is star dust.

- A **more massive star** cools and expands to become a **red supergiant**. The core will contract and fusion in the core forms elements with larger nuclei. When the nuclear fusion reactions are finished the star cools and contracts rapidly. The outer layers rebound violently against the dense core of the star. This is a **supernova**. It releases an enormous amount of energy. A supernova can be as bright as an entire galaxy. All the elements heavier than iron that exist naturally on planets were created in supernova explosions. The core is left as a **neutron star**. It has a large mass and is very dense.
- **Black holes** are the most dense neutron stars. They are so dense that even light cannot escape from their strong gravitational fields. This means we cannot see them. We can work out where they are because they attract gases from nearby stars. Matter accelerated towards the black hole gives out X-rays, which we can see.

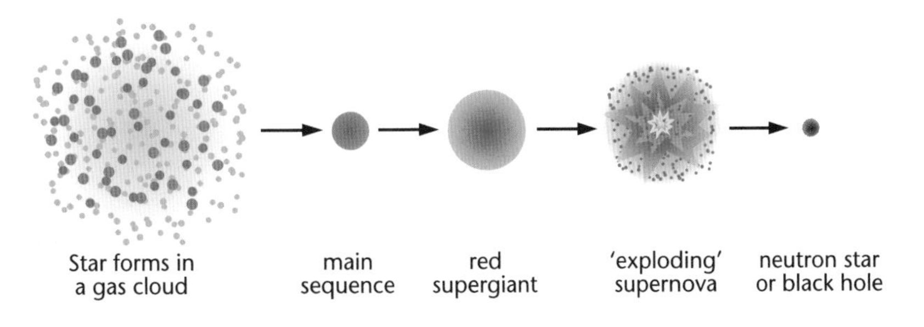

Fig. 5.9 The life cycle of a more massive star.

Galaxies and the Universe

OCR A P1.3
OCR B P2h
AQA P3.13.10
EDEXCEL 360 P1b12

Our Sun is a star in the **Milky Way galaxy**.
Galaxies are:

- collections of thousands of millions of stars
- all moving away from us, and from each other.

The Milky Way galaxy is shaped like a flat disc. There are thousands of millions of galaxies in the **Universe**. Between them there are empty regions that would take light hundreds of millions of years to cross.

The Universe is everything that exists: there is nothing outside the Universe – not even empty space.

5.5 Observing and exploring

Size: distance and time

OCR A P1.1/3
OCR B P2f
EDEXCEL 360 P1b12

The light year

Light travels through a vacuum at 300 000 km/s. Light from the Sun spreads out on a journey in all directions. The small amount we see on Earth arrives after about 10 minutes. After a **year**, **light** from the Sun will have **travelled** a distance of 300 000 km/s times the number of seconds in a year.

KEY POINT

A light year is the distance light travels in a year. Distances to stars and galaxies are so large that we measure them in light years. The nearest star to Earth is the Sun. The second nearest is about 4 light years from Earth.

A light year is about 9.5×10^{12} km, in words, nine and a half million million kilometres. You don't need to remember this number, but you need to understand how it is worked out.

When you look through a telescope and see a star 100 light years away, what you see – the light entering your telescope – **left** the star **100 years ago**. So looking at very distant planets is like **looking back** in **time**. **Distant** objects look **younger** than they really are.

If an alien 950 light years away looks at Earth through its telescope on the right day in 2016 it will be able to watch the Battle of Hastings in 1066.

Object	Details of size	Details of age
asteroid	wide range, from pebble to 1000 km across	
the Moon	smaller than Earth (diameter about a quarter of the Earth's)	4500 million years
dwarf planet Pluto	smaller than the Moon	
Earth	diameter 12 760 km	5000 million years
largest planet Jupiter	diameter over 10 times the Earth's	4600 million years
the Sun	diameter over 100 times the Earth's	
Solar System	diameter about ten thousand million km (10 000 000 000 km) (light would take between 10 and 11 hours to cross the Solar System)	
nearest star	about four light years away	
Milky Way galaxy	diameter about 100 000 light years	
the observable Universe	about 14 thousand million light years	About 14 thousand million years

Difficult observations

OCR A P1.1/3,P7.3

All the **information** we have about objects outside the Solar system comes from **observations** made with **telescopes**. What we know depends on the **electromagnetic radiation** from the stars and galaxies.

It is very difficult to work out how far away a star or galaxy is. A very bright star may look bright because:

- it is larger than other stars
- it is hotter than other stars
- it is closer than other stars.

If the colour of two stars is the same and there are reasons to believe they are similar stars, then a difference in brightness can be used to measure the distance. The **further away** the star, the **dimmer** it is.

Parallax

You may have noticed this effect from a train or car window; objects close to you seem to move more quickly, and change position, when compared to distant objects.

Another method is to use **parallax measurement**. The Earth orbits the Sun. If an **observation** of the night sky is made, and then we wait for **six months**, the Earth will have **changed** its **position** in space by a distance equal to the **diameter** of its **orbit**.

When the night sky is observed from this position, a **star** that is **close to Earth** will have **changed its position** when **compared** with more **distant stars** in the **background**. The amount the near star has moved can be used to calculate how far it is from the Earth. This is shown in Fig. 5.10.

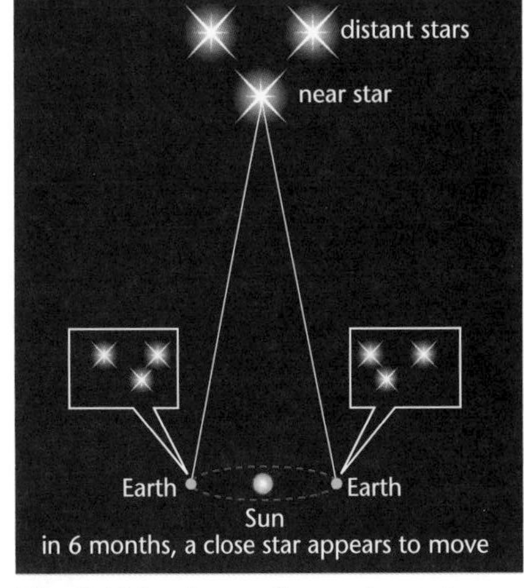

Fig. 5.10 In six months, a close star appears to move.

These difficulties in making observations lead to **uncertainty** in our **measurements** of the distances to stars and galaxies.

Light pollution

Another difficulty for astronomers is the amount of **light pollution** from the Earth. All the light from our cities shines into the night sky, which makes it difficult to pick up the very **weak signals** from **distant stars** and **galaxies**. To see the Milky Way (it looks like a milky strip of stars in the sky), you need to be far away from towns and cities.

Space travel in the Solar System

Manned missions

We have used **spacecraft** to put men and women in **orbit** and land men on the **Moon** and bring them home. There are plans to send humans to **Mars** but the difficulties of space missions to other planets are extreme:

- Enough **food**, **water**, **oxygen** and **fuel** must be carried for the entire trip, including any delays. The distances involved mean that emergency supplies could not be sent in time to be of use. An **artificial atmosphere** must be set up inside the spacecraft, and levels of carbon dioxide and oxygen have to be monitored.
- Interplanetary space is very cold. There must be **heating** to keep astronauts warm. As the spacecraft will receive direct radiation from the Sun, it will heat up, so **cooling** is required.
- There will be very **low gravity** during the journey. This will affect health. Bones lose density and muscles waste away under these conditions. A special exercise programme will be needed using **exercise machines**. Experiments have been done in creating artificial gravity by spinning spacecraft, but have not been very successful.
- Some way of **shielding** astronauts from **cosmic rays** is needed. The radiation released when there is a solar flare is deadly (see 5.1 Solar flares).
- **Distances** are so large that only the closest planets could be reached.

Unmanned missions

Unmanned spacecraft have some advantages. They can operate unharmed in a lot of conditions that would kill humans. By using **remote sensors** and computers they can send back information on:

- temperature
- magnetic field
- radiation
- gravity
- gases in the atmosphere
- composition of the rocks
- appearance of the surroundings (using TV cameras).

The **costs** of unmanned missions are **less** and there are **no lives at risk** if something goes wrong. The NASA 'Viking' spacecraft and the 'Spirit' and

'Opportunity' Mars rovers have all studied the rocks and terrain of Mars. We have a lot of information about Mars without any humans travelling there. The disadvantages are that everything must be thought of before the mission leaves – **no adjustments**, **repairs** or changes can be made unless they are programmed into the computers and can be done remotely. Most people are not as interested in unmanned missions, and they do not inspire people in the way that manned missions do.

Using telescopes

Beyond the Solar System **telescopes** are the only way to study the Universe. Even in the Solar System we have obtained a lot of information from telescopes. Telescopes can be positioned **on Earth** where they are **easier** to **maintain** and repair, but where people must look **through** the **atmosphere**. Some telescopes are positioned **high up** on mountain tops to get **above** the clouds, dust and pollution, and away from city lights. Telescopes in **orbit** avoid these problems and have a much clearer view of the stars. **Launching** a telescope into orbit is **expensive**.

There are **different telescopes** to use **all** the different **ranges** of the **electromagnetic spectrum**, including **radio telescopes**, which observe radio waves from space. As X-rays do not pass through the atmosphere, **X-ray telescopes** are always placed in orbit.

Life elsewhere in the Universe

Astronomers have identified some distant **stars** that have **planets** around them. These planets could be home to **alien life**. We have **not** yet discovered any trace of **alien life**, either existing now or that lived in the past. Even if only a few stars have planets, because there are so many stars, scientists think it **likely** that life has **evolved somewhere else** in the Universe. A life-form like ours would need a **planet close** enough to its **Sun** to be warm but not so close it would burn. The alien Sun would need to be in a stable part of its life – which probably means a **main sequence star**.

Even if we discovered life elsewhere it is difficult to see how we could **make contact** with the large distances and **time delays** for signals – any signal we receive from a million light years away was sent a million years ago.
The **Search for Extra-Terrestrial Intelligence** (SETI) is a project that scans **radio waves** from space for evidence of **patterns** that suggest aliens using radio waves for communication – as we do.

5.6 The expanding Universe

The Big Bang

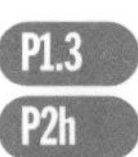

OCR A P1.3
OCR B P2h
AQA P1.11.7
EDEXCEL 360 P1b12

The **Universe** began with a **'Big Bang'** about **14 thousand million years ago**. It is still **expanding**. It is difficult to predict whether it will continue to expand, stop, or start to contract.

For Edexcel you need to know about the alternative theory to the Big Bang described in the How Science Works section on page 77.

KEY POINT

Hubble's Law:

The recessional speed of a galaxy is directly proportional to its distance from our galaxy. The further away a galaxy is, the faster it is moving away from our galaxy.

Hubble's Law is strong evidence for the Big Bang because it suggests that the Universe started expanding from a small starting point.

Microwave background radiation

OCR B P2h
EDEXCEL 360 P1b12

Microwave background radiation comes from all parts of the **Universe** and is the left over radiation from the **Big Bang**. It has **cooled** to the **microwave** region of the electromagnetic spectrum.

Do not confuse this with radioactive background.

All the information we have about distant stars and galaxies comes from the radiation they produce.

Red shift

OCR B P2h
AQA P1.11.7
EDEXCEL 360 P1b12

If a **source of waves** is moving **away**, the **wavelength** appears **longer** and the **frequency** of the waves appears **lower**. Fig. 5.11 shows how in the time between one crest and the next the source moves further away so the wavelength is longer than for the stationary source.

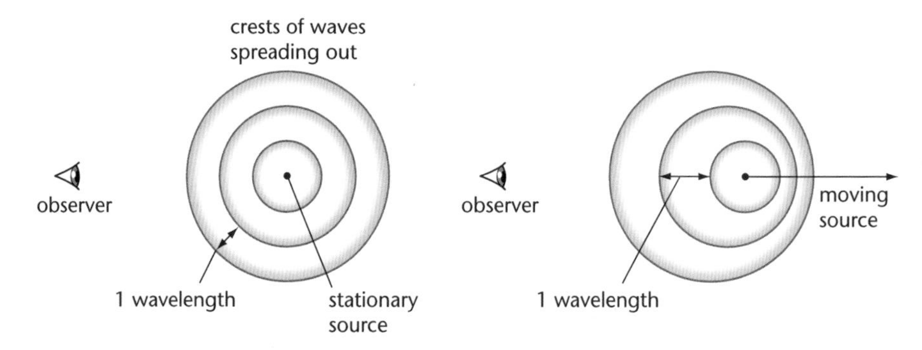

Fig. 5.11 Waves from a moving source.

If a source of light waves is moving away so that the wavelength is **longer**, this is called a **red shift**.

A red shift in the light from a star shows that the distance between us and the star is increasing. The **bigger** the **red shift**, the **faster** the star is **moving away**.

Scientists think that the red shift we see for all galaxies is because space is expanding, not because we, or the star, are moving through space.

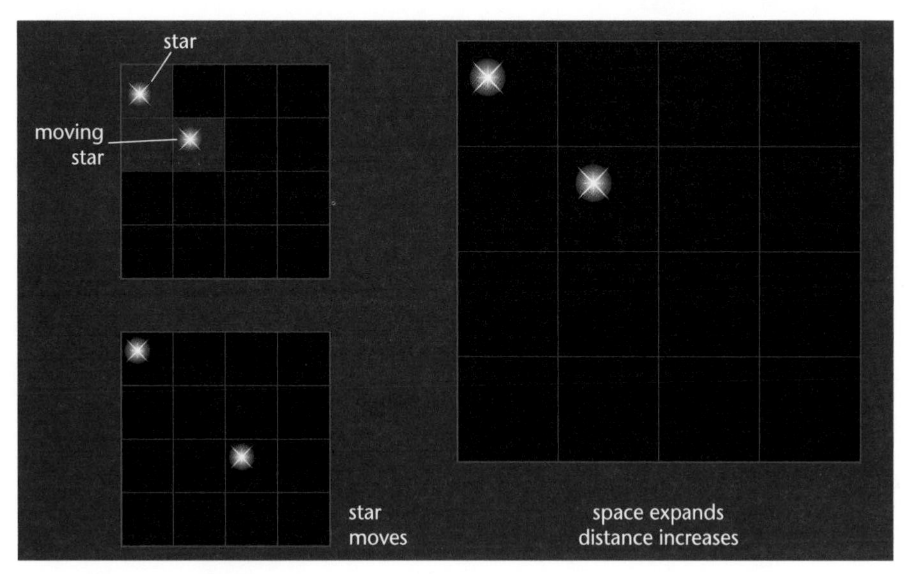

Fig. 5.12 The difference between a moving star and expanding space.

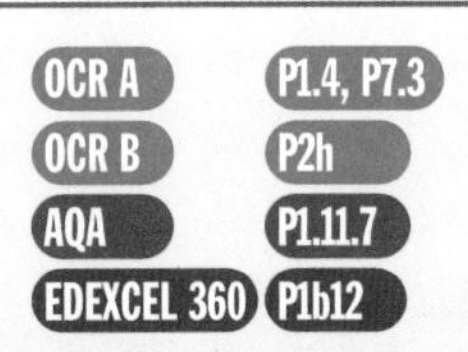

HOW SCIENCE WORKS

Ideas about the Universe

In 1920, there was a 'great debate' between two astronomers. Each gave a talk about ideas that some astronomers had at the time.

Harlow Shapley 1920

"Our Sun is not at the centre of our galaxy. Our galaxy is very big."

"There is only one galaxy."

or

Herber Curtis

"The fuzzy objects in the sky known as 'spiral nebulae' are not gas clouds but other galaxies at great distances from our galaxy."

"The Sun is near the centre of our small galaxy."

At the time there was no evidence to prove who was right or who was wrong.

Evidence: in 1924, Edwin Hubble showed that, as Curtis thought, the distance to 'spiral nebulae' was much greater than the size of the Milky Way. In the 1920s and 1930s, the size of our galaxy was measured more accurately, showing, as Shapley thought, it was large and the Sun was not central.

Edwin Hubble also used the red shift to show that all galaxies are moving away from all other galaxies. Those furthest away were moving fastest.

New ideas

George Gamow 1946

"If all galaxies are moving away from each other, then at some time in the past they must all have started from the same point. This 'explosion' was the beginning of the Universe. The 'explosion' would produce radiation. By now it will have cooled to microwaves."

HOW SCIENCE WORKS

Fred Hoyle 1950

“Steady state theory makes more sense. The Universe must always have existed. I prefer the steady state theory to an ‘explosion’: it’s just a ‘Big Bang’ theory.”

Evidence: Astronomers started to look for the microwave radiation. In 1965, Arno Penzias and Robert Wilson discovered that a low level of microwave radiation came from all directions in the Universe. They were not looking for the radiation and did not know what it was. They had found the cosmic microwave background which supports the Big Bang theory.

The next new ideas

“The expansion of the Universe will slow down, and it will contract to a point, because of the gravitational attraction between all the mass in the Universe. There is enough mass in the Universe for this to happen.”

“After the Universe has contracted to a point there will be a new Big Bang. This is called the oscillating Universe theory.”

“The Universe will continue to expand forever, because there is not enough mass to provide the gravity to reverse the expansion.”

“There is just enough mass to stop the expansion of the Universe, but not to start it contracting again.”

The evidence scientists need is how much mass there is in the Universe. They know there is some mass they cannot see. This is because objects they can see are reacting to larger gravitational forces caused by the mass they cannot see. Scientists call this Dark Matter.

What is Dark Matter **?** *How much is there* **?**

Scientists are searching for evidence to support the different ideas.

1. (a) What two pieces of evidence are there for the Universe starting with a Big Bang?
   (b) What ideas do scientists have about what will happen to the Universe in the future?
   (c) What will decide which of these ideas will happen?
   (d) Explain whether it is possible for scientists to find the answer to what happened before the Big Bang. **[6]**
2. Put these statements in order to explain the steps in developing the Big Bang theory of the Universe.
   A A scientist suggests an explanation.
   B Scientists look for evidence that the predictions are true.
   C A scientist makes some observations.
   D Scientists find the evidence and the theory is accepted.
   E Scientists use this to make some predictions. **[4]**

Exam practice questions

1. Put these objects in order of size, starting with the smallest.

Earth **Milky Way Galaxy** **Solar System** **Sun** **Universe** **[1]**

2. Which of these best describes the way galaxies move?

A Galaxies are stationary, but our galaxy is moving away from them.
B Galaxies are all moving away from each other at constant speed.
C Galaxies are all moving away from each other. Those furthest away are moving fastest.
D Some galaxies are moving away from our galaxy (the Milky Way). **[1]**

3. Which of these is **not** evidence for the theory that the continents of Europe and North America are drifting apart?

A There are similar fossils in the two continents but different modern animals.
B The shape of the continents suggests they fit like pieces of a jigsaw.
C The rocks at either side of the Mid-Atlantic ridge are in strips that are magnetised in opposite directions.
D Mountains like the Himalayas are formed when a continental plate collides with another continental plate. **[1]**

4. Use words from the list to complete this description of the life of a star like our Sun.

black dwarf **contracts** **fusion** **helium** **hydrogen** **main sequence**
nebula **nuclear** **planetary nebula** **protostar** **red giant** **white dwarf**

A large cloud of dust and gases called a _____1_____ or an interstellar gas cloud, _____2_____ and heats up forming a _____3_____. In a new star the process called _____4_____ _____5_____ starts. This is when _____6_____ nuclei join to form _____7_____ nuclei. The star will be a _____8_____ star for a long time.

When all the _____9_____ has been fused the star becomes a _____10_____. As it cools the outer layers move away as a _____11_____ leaving a _____12_____ which will eventually cool to a _____13_____ **[9]**

5. The following statements about space travel are reported in the press. Write **T** for the statements that are **true** and **F** for the statements that are **false**.

(a) A manned space trip to the nearest star would take four years.
(b) People living on the Moon would experience no gravity, so they would lose bone mass and their muscles would waste away.
(c) A spacecraft would need to be shielded from cosmic rays as the radiation is dangerous for humans.
(d) Unmanned spacecraft can use remote sensors to send back information about temperature, the magnetic field, gases in the atmosphere, and photographs.
(e) Unmanned missions to Mars have sent back a lot of information about Mars without humans travelling there.
(f) A spaceship will require heating to keep astronauts warm. **[3]**

Chapter

6 Forces and motion

The following topics are covered in this chapter:

- *Speed, velocity and acceleration*
- *Momentum and collisions*
- *Forces and their effects*
- *Work, energy and power*

6.1 Speed, velocity and acceleration

Speed and distance-time graphs

OCR A P4.1, OCR B P3a, AQA P2.12.1

Speed is measured in metres per second (m/s) or kilometres per hour (km/h). If an athlete runs with a **speed** of 5 m/s, she will cover 5 metres in one second and 10 metres in two seconds. An athlete with a faster speed of 8 m/s will travel further, 8 m in each second, and will take less time to complete his journey.

For how to rearrange the formula see page 225.

To calculate speed:

$$\textbf{speed (m/s)} = \frac{\textbf{distance (m)}}{\textbf{time (s)}}$$

Direction of travel

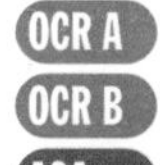

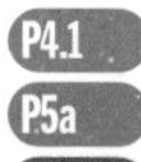

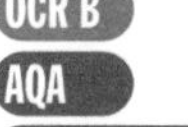

OCR A P4.1, OCR B P5a, AQA P2.12.1, EDEXCEL 360 P2.9

There are two ways of looking at a journey:

- You can say that the **distance** you travel can only increase or stay the same, and then the **speed** is always a positive number.
- You can consider the **direction** you travel, so that if you travel towards school, that is a **positive** distance and when you travel in the opposite direction that is a **negative** distance. Sometimes, distance in a given direction is called **displacement**.

You only need to know the term 'displacement' for Edexcel.

Quantities that have a **magnitude** and **direction** are called **vectors**.

Velocity is a **vector**, because velocity is speed in a given direction.

Example A boy walks in a positive direction and then back again with a constant **speed** of 2 m/s, so he walks with a **velocity** of +2 m/s and then with a velocity of –2m/s.

For how to rearrange the formula see page 225.

$$\textbf{velocity} \text{ (m/s)} = \frac{\textbf{displacement} \text{ (m)}}{\textbf{time} \text{ (s)}}$$

Distance–time graphs

OCR A P4.1
OCR B P3a
AQA P2.12.1

On a **distance–time graph**:

- a **horizontal line** means the object is **stopped**
- a **straight line sloping upwards** means it has a **steady speed**.

The steepness, or **gradient**, of the line shows the speed:

- a **steeper gradient** means a **higher speed**
- a **curved line** means the **speed is changing**.

If the direction of travel is being considered:

- A negative distance is in the opposite direction to a positive distance.
- A **straight line sloping downwards** means it has a **steady speed**, and a **steady velocity** in the negative direction.

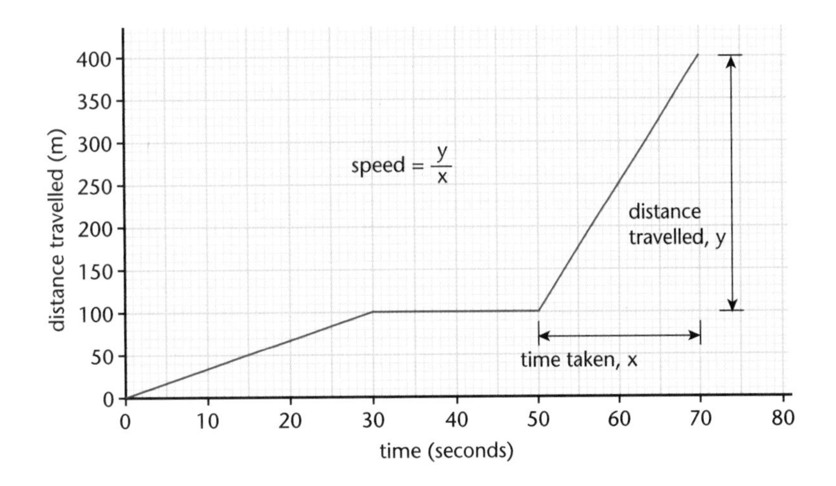

Fig. 6.1 Distance–time graph for a cycle ride.

Between 30 s and 50 s the cyclist stopped. The graph has a **steeper gradient** between 50 s and 70 s than between 0 s and 20 s – the cyclist was travelling at a **greater speed**.

To calculate a speed from a graph, work out the gradient of the straight line section as shown in Fig. 6.1:

speed = $\frac{y}{x}$ where y = 400 m – 100 m = 300 m and x = 70 s – 50 s = 20 s.

$$\text{speed} = \frac{300\text{ m}}{20\text{ s}} = 15\text{ m/s}.$$

Average speed and instantaneous speed

You can calculate the **average speed** of the cyclist for the **total** journey in Fig. 6.1 using:

$$\text{average speed} = \frac{\text{total distance}}{\text{total time}}$$

$$\frac{400\text{ m}}{70\text{ s}} = 5.7\text{ m/s}$$

This is an important point for OCR A.

This is not the same as the **instantaneous speed** at any moment because the speed changes during the journey. If you calculate the **average speed** over a **shorter time** interval you get closer to the **instantaneous speed**.

Velocity–time and speed–time graphs

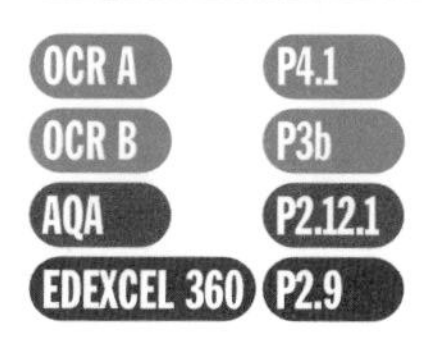

A **change of velocity** is called **acceleration**. Speeding up, slowing down and changing direction are all examples of acceleration.

Fig. 6.2 shows how to interpret a velocity–time graph.

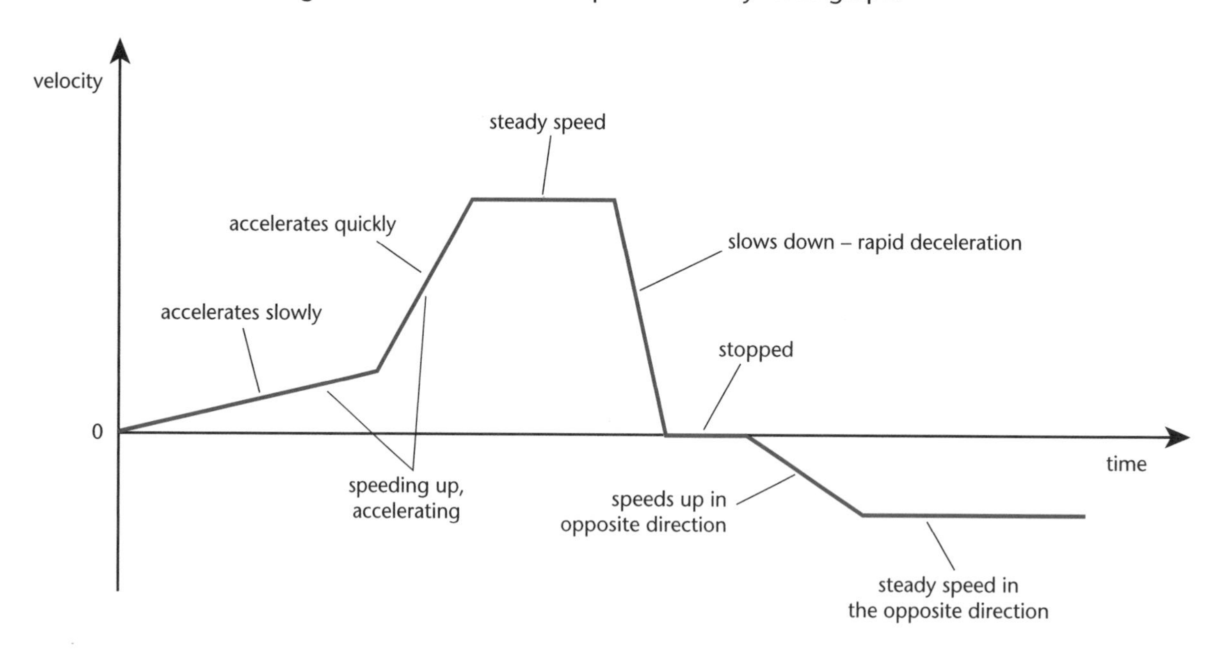

Fig. 6.2 A velocity–time graph.

- A **positive slope (gradient)** means that the **speed is increasing** – the object is **accelerating**.
- A **horizontal line** means that the object is travelling at a **steady speed**.
- A **negative slope (gradient)** means the **speed is decreasing – negative acceleration**.
- A **curved slope** means that the **acceleration** is **changing** – the object has **non-uniform acceleration**.

Check carefully whether a graph is a speed-time graph or a distance-time graph.

On true **speed–time graphs**, the speed has only positive values. On **velocity–time graphs** the velocity can be negative.

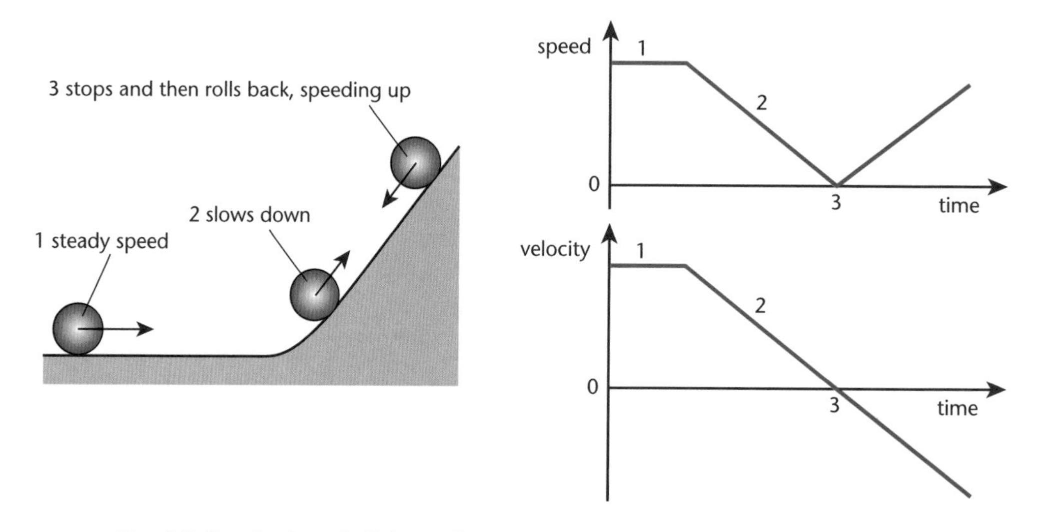

Fig. 6.3 Graphs for a ball that rolls up a hill, slows, and rolls back down, speeding up.

Tachographs are instruments that are put in lorry cabs to check that the lorry has not exceeded the speed limit, and that the driver has stopped for breaks. They draw a graph of the speed against time for the lorry.

Graphs, acceleration and distance

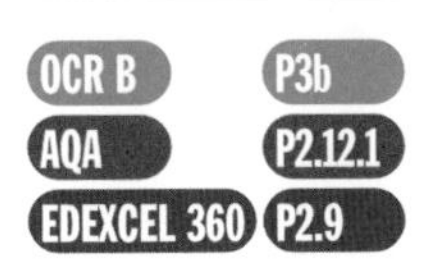

For how to rearrange the formula see page 225.

KEY POINT

Acceleration is the change in velocity per second:

$$\textbf{Acceleration (m/s}^2\textbf{)} = \frac{\textbf{change in velocity (m)}}{\textbf{time taken (s)}}$$

$a = \dfrac{(v - u)}{t}$ **where a is the acceleration of an object whose velocity changes from u to v in time t**

Fig. 6.4 is a graph of speed against time for a car journey.

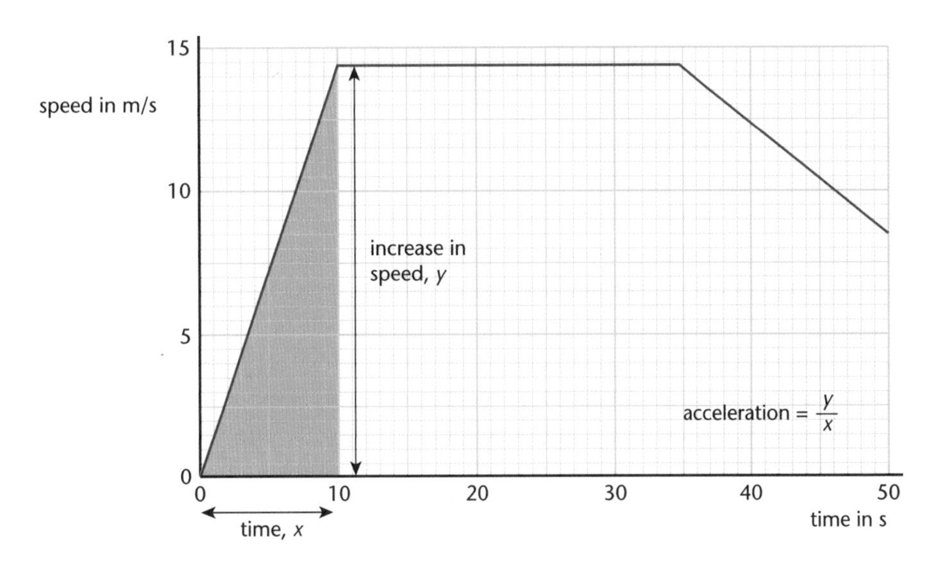

Fig 6.4 Speed–time graph for a car journey.

In the first 10 s the acceleration is: $\dfrac{(14 - 0)\text{ m/s}}{10\text{ s}} = 1.4\text{ m/s}^2$

The distance travelled in the first 10 seconds is given by the shaded area.

Distance travelled = average speed × time

average speed = $\frac{1}{2}$ (14 – 0) m/s and time = 10 s

Distance travelled in metres = $\frac{1}{2}$ (14 – 0) × 10 which is the area of the shaded triangle ($\frac{1}{2}$ base × height).

The distance travelled in the next 25 s is represented by the rectangle.

Distance = 14 m/s × 25 s = 350 m

On a speed–time graph, the area between the graph and the time axis represents the distance travelled.

6.2 Forces and their effects

Mass and weight

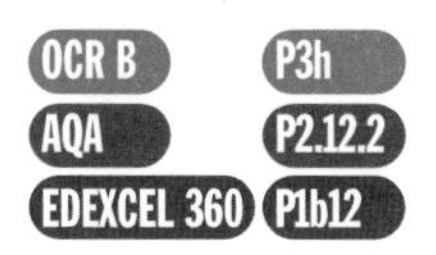

Mass is measured in **kilograms**. An object has the same mass on the Earth, on the Moon, or far out in deep space. **Weight** is a **force** and is measured in **newtons**. Weight is due to **gravity** attracting the mass towards the centre of the Earth. In deep space there is no gravity and the mass has no weight. On the Moon, which has less mass than the Earth, the gravitational attraction is less, so objects will have less weight than they have on the Earth (only one sixth).

For how to rearrange the formula see page 225.

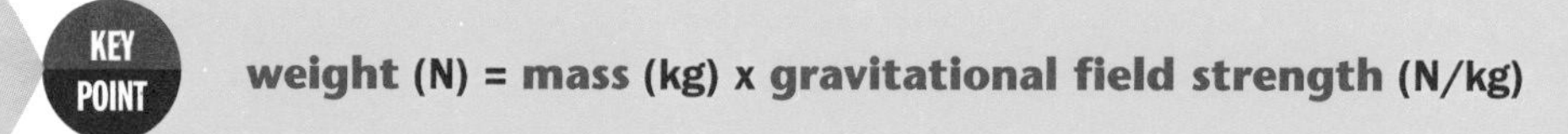

Gravitational field strength close to the surface of the Earth is assumed to be a constant 10 N/kg. Objects in free fall have a constant acceleration of 10 m/s^2.

The resultant force and balanced forces

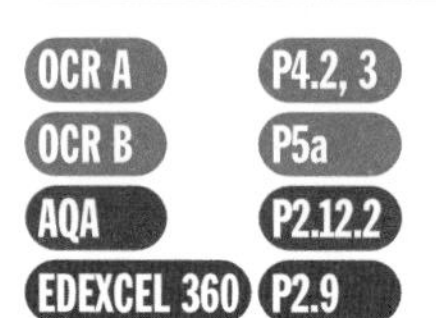

Forces have **size** and **direction**. On diagrams they are represented by arrows, in the **direction** the force acts. The **length** of the arrow represents the **size** of the force.

When an object has several forces acting on it, the effect is the same as one force in a certain direction. This is called the **resultant force**. Fig. 6.5 shows how forces can be combined to give a resultant force. If the **resultant force** is **zero** the forces on the object are **balanced**.

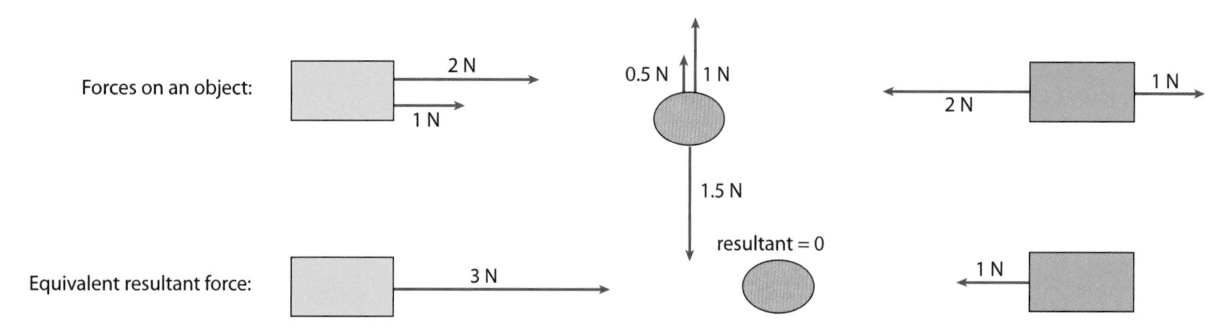

Fig. 6.5 resultant forces.

A **resultant force** is needed to change the **velocity** of an object. If the **forces** on an object are **balanced** then it will remain **stationary**, or, if it is moving, it will continue to move at a **steady speed** in the **same direction**. This seems strange because we are used to frictional forces slowing things down.

If the resultant force on an object is zero, the object will remain stationary or continue to move at a steady speed in the same direction.

Fig. 6.6 shows how the forces on an object can be balanced so that it does not fall.

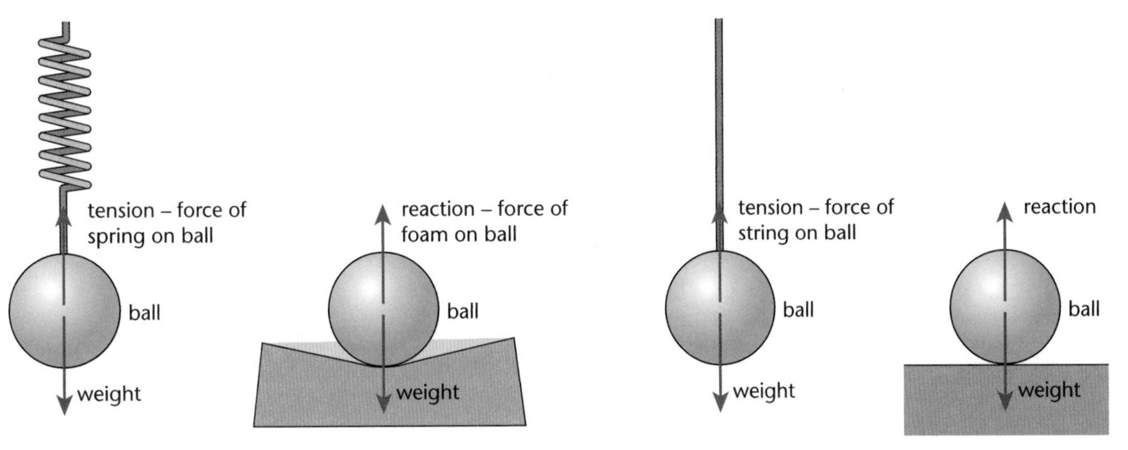

Fig. 6.6 Reaction and tension are forces that can balance weight.

Resistance to motion – friction and drag

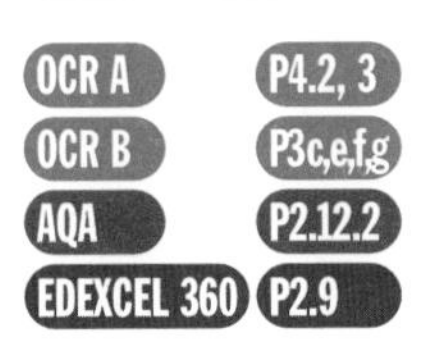

When one object slides over another there is **friction**, a **resistive force** between the two surfaces. This arises because, on a microscopic scale, the surfaces are not completely smooth and the high points become stuck together.

Air resistance (or **drag**), is a **resistive force** that acts against objects that are moving through the air. **Drag** acts on objects moving through any **fluid** (gas or liquid) – and is larger in liquids.

Friction and **drag forces**:

- always act **against** the direction of motion
- are **zero** when there is **no movement**
- increase as the **speed** of the object increases.

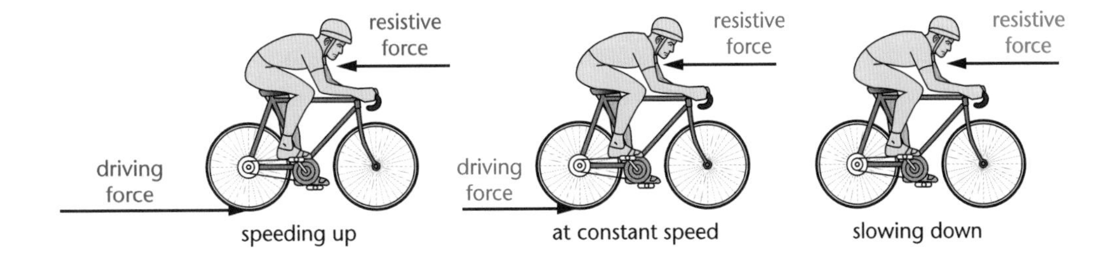

Fig. 6.7 The forces on a cyclist when speeding up, at constant speed, and slowing down.

When the driving force is **larger** than the resistive force, the cyclist **speeds up**; when they are **equal** he travels at a **steady speed**, and when the driving force is less than the resistive force he **slows down**.

When an object falls it can reach a **steady speed** called **terminal velocity** where the **drag** equals the **weight**,

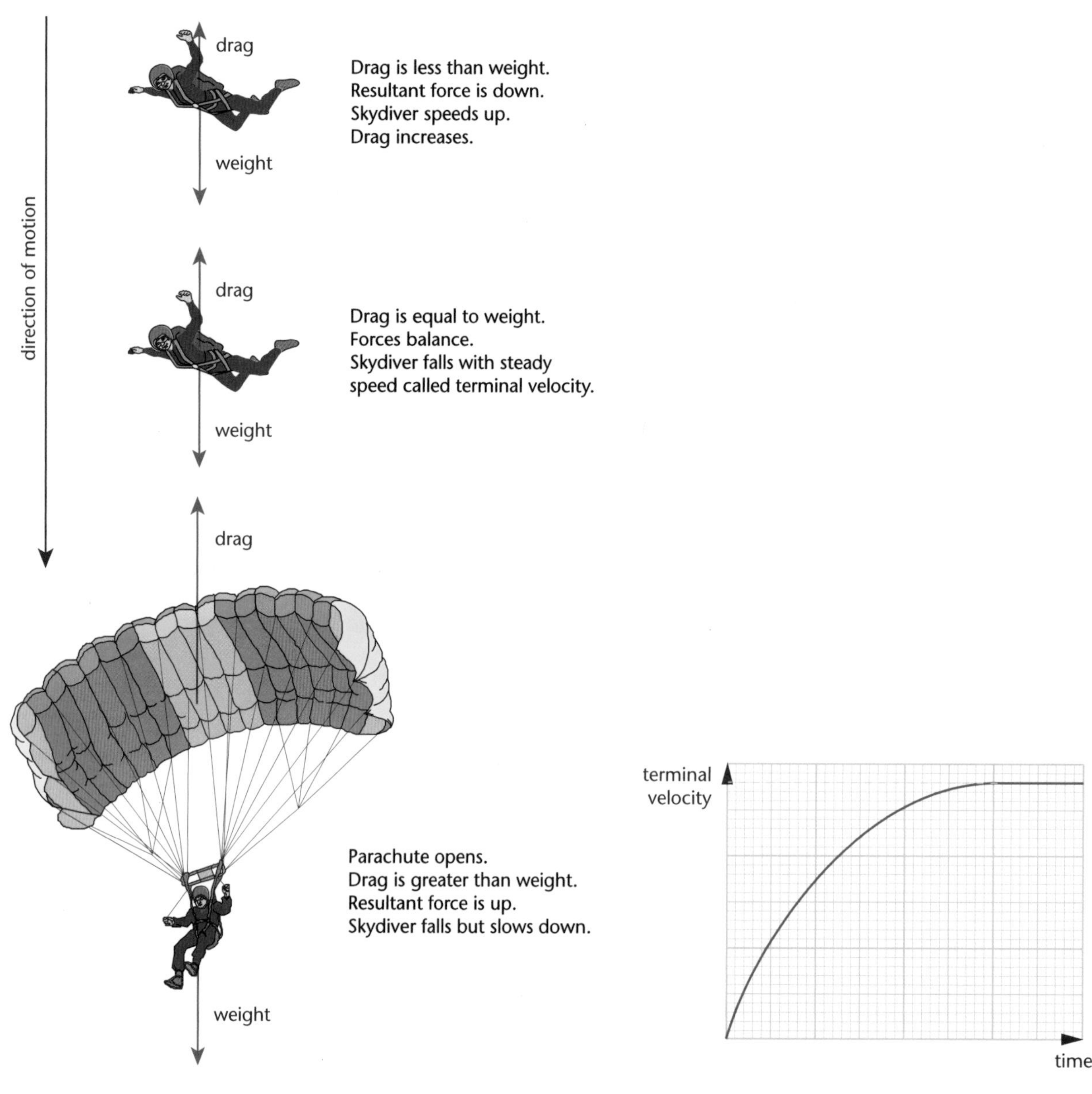

Fig 6.8 Forces when skydiving.

Fig 6.9 A falling object eventually reaches a terminal velocity.

Interaction pairs – action and reaction forces

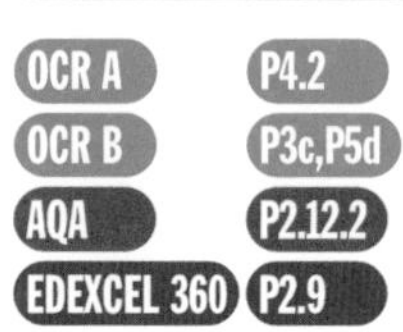

When two objects interact, there is always an **interaction pair** of **forces**. In Fig. 6.10, skater A cannot push skater B without skater B pushing skater A.

In an **interaction pair of forces**, the two forces:

- are always **equal** in size and **opposite** in direction
- always act on **different** objects.

Fig. 6.10 when two skaters push each other, they both move backwards.

Do not confuse these forces, which act on different objects, with balanced forces, which act on the same object.

When two objects interact, the forces they exert on each other are equal and opposite and are called action and reaction forces.

A **rocket** and a **jet engine** make use of this effect. There is an equal and opposite **reaction force** to the force of the hot exhaust gases pushed out of the back. This reaction force sends the rocket or jet forward.

Friction is the force that provides the **reaction force** you need for walking or for transport using wheels. Fig. 6.11 shows the force of **the wheel pushing back on the road** and the reaction – the friction force – of **the road pushing forward on the wheel**, which sends the wheel forward.

Imagine walking, or cycling, on a frictionless icy surface. You would slip back and never move forward.

Fig. 6.11 Forces that make a wheel move forward.

Descriptions of what is pushing/pulling on what are important in OCR A.

When you walk, **your foot pushes back on the ground** and the **ground pushes forward on your foot**.

Force and acceleration

P2.12.2

EDEXCEL 360 P2.9

A **resultant force** on an object causes it to **accelerate**. The acceleration:

- is in the same **direction** as the force
- increases as the size of the **force** increases
- depends on the **mass** of the object – is smaller for a larger mass.

For how to rearrange the formula see page 225.

For a resultant force on an object with mass m:

Force (N) = mass (kg) × acceleration (ms^{-2})

Stopping distances

The distance that a vehicle travels between the driver noticing a hazard and when the vehicle stops is called the **stopping distance**:

stopping distance = thinking distance + braking distance

- **Thinking distance** is the distance travelled during the driver's reaction time – the time between seeing the hazard and applying the brakes.
- **Braking distance** is the distance travelled while the vehicle is braking.

This diagram shows the shortest stopping distances at different speeds.

When speed doubles, thinking distance doubles, and braking distance is four times as far.

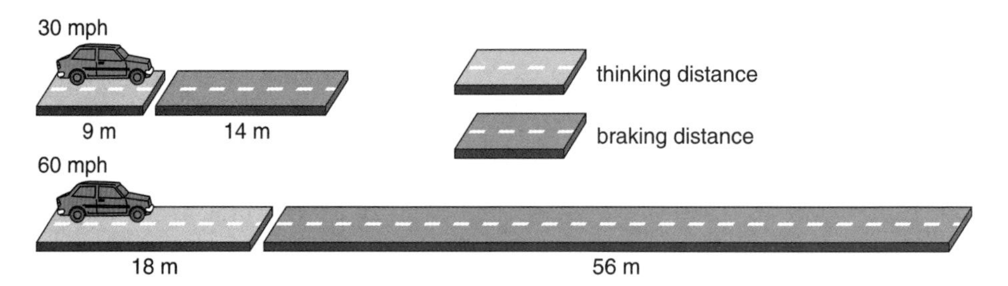

Fig. 6.12 The stopping distance increases with speed.

The **stopping distances** will also be longer if:

- The **driver** is **tired**, or affected by **drugs** (including **alcohol** and some medicines,) or **distracted** and not concentrating. Reaction time, and thinking distance are increased.
- The **road** is **wet** or **icy** or the **tyres** or **brakes** are in poor condition. The friction forces will be less, so the braking distance will increase.
- The **vehicle** is **fully loaded** with passengers or goods. The extra **mass** reduces the deceleration during braking, so the braking distance is increased.

These stopping distances are taken into account when road speed limits are set. Drivers should not drive closer than the thinking distance to the car in front, allowing them time to react. They should also reduce speed in bad weather to allow for the increased braking distance.

An anti-lock braking system (ABS) helps to keep the stopping times to a minimum. If the wheels start to slip because there is not enough friction (making the stopping distance longer), the system disconnects the brakes for a moment, the wheel grips the road, and the system reapplies the brakes.

6.3 Momentum and collisions

Momentum

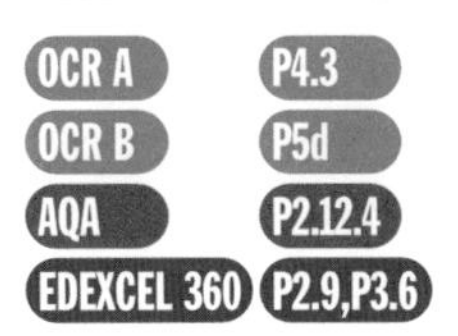

Ten-pin bowling balls are available with different **mass**. A fast moving ball with low mass can be as effective as a slow ball with high mass. This is because it is the **momentum** of the ball that is important in the **collision** with the pins, and **momentum** depends on **mass** and **velocity**.

For how to rearrange the formula see page 225.

KEY POINT

momentum (kg m/s or Ns) = **mass** (kg) × **velocity** (m/s)

Momentum is a vector with the same direction as the velocity. The unit kg m/s is the same as the N s.

When a **force** acts on an object that is moving, or able to move, it causes a **change in momentum** in the same direction as the force (because the velocity changes). The longer **time** that the force acts, the bigger the change in momentum.

KEY POINT

change in momentum (kg m/s or Ns) = **force** (N) × **time** (s)

The time is the time for which the force acts:

$$\text{force (N)} = \frac{\text{change in momentum (kg m/s or Ns)}}{\text{time (s)}}$$

When two objects **collide** or **explode** apart, there is an equal and opposite force on each object, and they interact – push against each other – for the same time. This means that the **change in the momentum** of the objects is equal and opposite. Another way to say this is that the total momentum of the two objects before the collision or explosion is the same as the total momentum after the collision.

KEY POINT

If no external forces act on the colliding/exploding objects, the total momentum of objects before a collision/explosion is the same as the total momentum after the collision/explosion.

This is called the conservation of momentum.

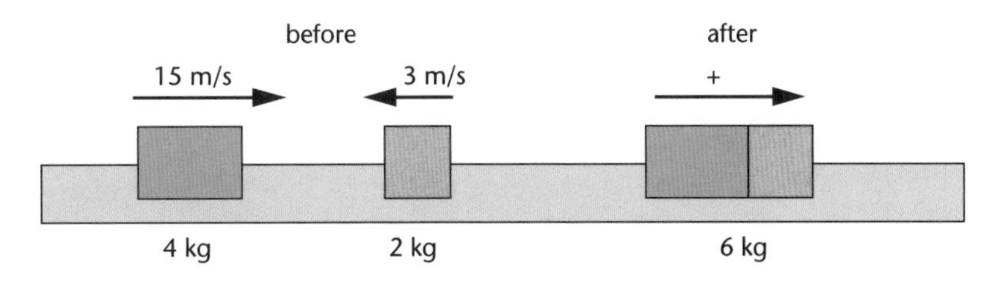

Fig. 6.13 Two objects colliding and sticking together.

Calculations using conservation of momentum are not required for Edexcel.

Taking the right direction to be positive:

Before the collision momentum =

$$4\,\text{kg} \times 15\,\text{m/s} + 2\,\text{kg} \times (-3)\text{m/s} = (60 - 6)\,\text{kg m/s}$$

After the collision, momentum = 6 kg × v

So, because of conservation of momentum, v = 9 m/s

Safer collisions

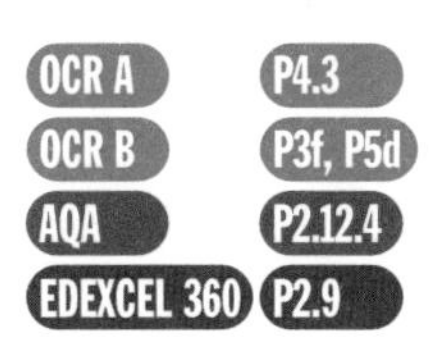

When your body is in a **collision**, a **force** brings it to a sudden stop. The larger the stopping force on the body, the more it is damaged. To reduce **damage** we must reduce the **force**. This means:

- reducing the acceleration (force = mass × acceleration) which means reducing the velocity of the body more slowly.

This the same as:

- reducing the momentum of the body more slowly.

$$(\text{force} = \frac{\text{change of momentum}}{\text{time}})$$

If the collision takes place over a longer time, say 0.5 s instead of 0.05 s – ten times as long – then the stopping force will only be one tenth of the size. The time of a collision can be increased by using:

- **Crumple zones** The car occupants are in a strong safety cage. The front and back of the car are designed to crumple in a collision, increasing the distance and time over which the occupants are brought to a stop.

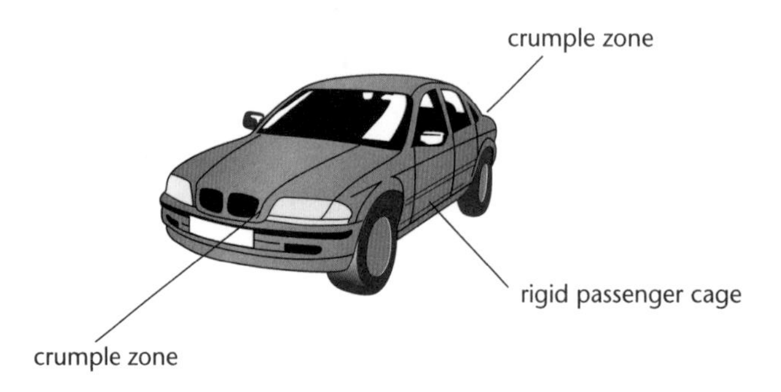

Fig. 6.14 Crumple zones

- **Airbags** The body hits the airbag, which is compressed, increasing the distance the body moves and the time it takes to stop.
- **Seatbelts** These are designed to stretch slightly so that the body moves forward and comes to a stop more slowly than it would if it hit the windscreen or front seats. After a collision, the seatbelts should be replaced because having stretched once, they may not work properly again.
- **Cycle and motorcycle helmets** These contain a layer of material which will compress on impact so that the skull is brought to a stop more slowly. They should be replaced after a collision as the material will be damaged and may not give good protection again.

6.4 Work, energy and power

Work and energy

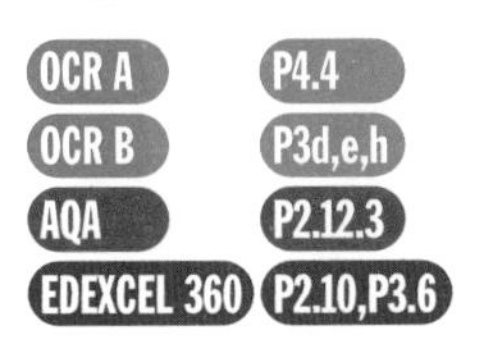

Whenever a **force** makes something move, **work** is done. The amount of **work done** is equal to the amount of **energy transferred**. **Work**, like **energy**, is measured in **joules**.

For how to rearrange the formula see page 225.

KEY POINT

work done by a force (J) = force (N) × distance moved by force in direction of the force (m)

When work is done by something, it loses energy; when work is done on something it gains energy.

Gravitational potential energy

A rollercoaster at the top of a slope has **stored energy**, which is called **gravitational potential energy (GPE)**, or sometimes potential energy (PE). This is the name used to describe stored energy that an object has because of its position – in this case, higher above the surface of the Earth.

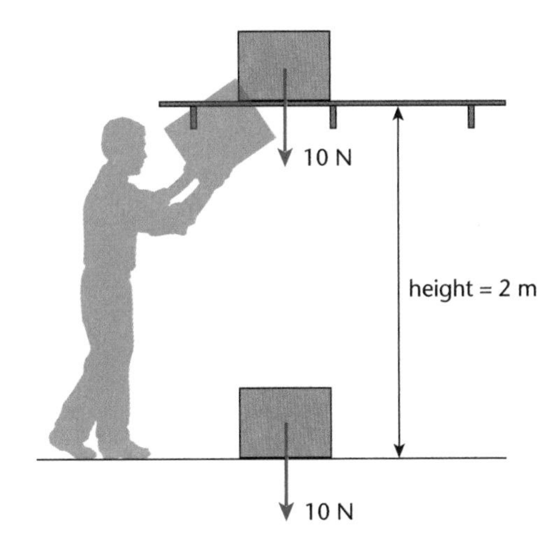

Fig. 6.15 Doing work – increasing the GPE.

Fig. 6.15 shows that when you lift a 10 N weight (a mass of 1 kg) from the floor to a high shelf, a height difference of 2 m, you have done work on the weight. The work done = 10 N × 2 m = 20 J and this is equal to the increase in the GPE of the weight.

Edexcel uses *g* = acceleration of free fall. This is the same as the gravitational field strength g = 10 m/s^2 = 10 N/kg and the units are equivalent.

KEY POINT

Change in gravitational potential energy, GPE (J) = weight (N) × vertical height difference (m)

or using *g* = the gravitational field strength (N/kg)

change in GPE (J) = *m* (kg) × *g* (N/kg) × *h* (m)

On Earth, the gravitational field strength is $g = 10$ N/kg but on the Moon, because it is smaller g is only about one sixth of this. The size of g increases with the mass of the planet or star. In exams, you will always be given a value to use for g.

Kinetic energy (KE)

This explains why stopping distances increase by four times when the speed is doubled – there is four times the kinetic energy to transfer by the braking force.

An object that is moving has **kinetic energy (KE)**. The energy depends on the **mass** of the object and on the **square** of the **speed** – doubling the speed gives four times the energy.

For example, an air hockey puck, floating on an air table, is almost frictionless. A **force** does **work** on the puck by pushing it a small **distance**. Energy is transferred increasing the **kinetic energy** of the puck – it speeds up. When the force stops the puck moves at a constant **speed** across the table – its **kinetic energy** is now constant.

For how to rearrange the formula see page 225.

KEY POINT

A mass that is moving has kinetic energy:

kinetic energy (J) = $\frac{1}{2}$ × mass (kg) × [speed (m/s)]^2

Or

kinetic energy (J) = $\frac{1}{2}$ × mass (kg) × [velocity (m/s)]^2

Energy does not have a direction. Speed or velocity can be used to calculate the kinetic energy.

Roller coasters and falling

When frictional forces are small enough to be ignored, the transfer of energy between **KE** and **GPE** can be used to calculate **heights** and **speeds**. Fig. 6.16 shows a car that is driven by a trackside motor to the top of a rollercoaster slope and then freewheels down the slope.

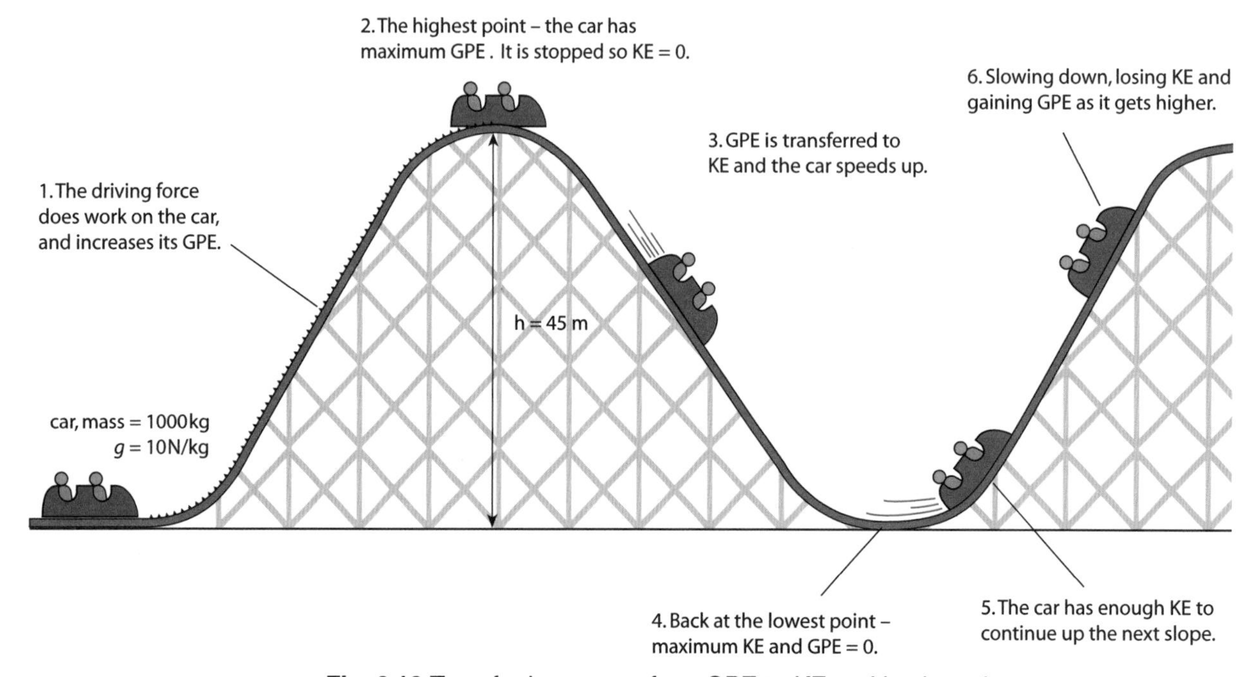

Fig. 6.16 Transferring energy from GPE to KE and back again.

Increase in GPE of train = $m\,g\,h$ = 1000 kg × 10 N/kg × 45 m = 450 000 J

Assuming there are no friction forces as the train travels down the slope:

loss of GPE = gain in KE

Remember this equation.

$$450\,000\text{ J} = \frac{1}{2}\,m\,v^2 = \frac{1}{2} \times 1000\text{ kg} \times v^2$$

$$v^2 = 900\,(\text{m/s})^2 \text{ so speed} = 30\text{ m/s}$$

Circular motion

EDEXCEL 360 P2.10

When a rollercoaster travels at a **constant speed** on a circular part of the track, it is **accelerating** – its velocity is changing because it is changing direction. There is a **resultant force** making the train move in a **circle**. The **resultant force** and the **acceleration** are directed **towards the centre** of the circle. This is always true for any object moving in a circular path.

In Fig. 6.17, a ball on a string moves in a circle. The **force** in this case is the **tension** in the string – cut the string and the ball flies off at a tangent – a straight line.

Fig. 6.17 A circular path needs a force towards the centre of the circle.

Work, friction and conservation of energy

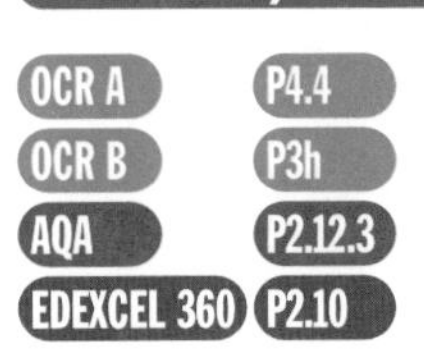

You can only use the relationship '**gain in KE = loss in GPE**' for a falling object if the drag (air resistance) is small and can be ignored – or if the object is falling in a vacuum. The skydiver in Fig. 6.8 in 6.2 Resistance to motion – friction and drag, eventually reaches **terminal velocity**. He is still falling, so GPE is being lost, but no KE is being gained. The energy is being used to do **work against** the **frictional force** (drag) and the skydiver and surrounding air will **heat up**.

For an object like the space shuttle, with a lot of **KE**, heatproof tiles are needed to protect it from the **heat** resulting from doing **work against** the **drag** when it re-enters the Earth's atmosphere.

Another example is a cyclist travelling at a **steady speed**. Energy is being transferred by **heating** the bicycle and surroundings as **work** is done **against friction**. No energy is being transferred as KE to the bicycle unless it speeds up.

For AQA, you need to know that some objects can return to their original shape because elastic potential energy is stored in the object when work is done to change its shape.

When **energy** is transferred to the surroundings by **heating** (for example due to frictional forces) it is no longer useful – but it is not lost. We say it has been **dissipated** (spread out) as **heat**. The **total energy** remains the same. This important result is called the **principle of conservation of energy**. Energy can be stored and transferred in different ways, but when it is all accounted for, the total amount stays the same.

Power

Power is the **rate of energy transfer**. This is the **work done** or **energy transferred** divided by **time**.

For how to rearrange the formula see page 225.

power (W) = $\frac{\text{work done or energy transferred (J)}}{\text{time (s)}}$

power is measured in watts (W) when energy is in joules (J) and the time in seconds (s)

Fuel consumption

Fuel consumption is measured in **litres per km**. (We also uses miles per gallon as a measure of performance – a high figure is good.)

Larger vehicles, and those with a greater acceleration have **more powerful engines**, so that energy can be transferred more quickly from the fuel to do work. The **work done = driving force** × **distance travelled**. A more **powerful** engine can transfer more energy per second – this uses **more fuel**.

Energy from fuel is needed to:
- increase KE
- do work against friction.

When a car brakes, the KE is transferred to heat. Braking and accelerating repeatedly will transfer more energy than travelling at a constant speed. The amount of braking and accelerating will depend on **driving style** and the **road conditions** (e.g. single carriageway, lots of bends, traffic lights). At **higher speeds**, resistive forces are greater and more work is done against friction. If more energy is transferred the fuel consumption is increased (increasing cost, and pollution).

HOW SCIENCE WORKS

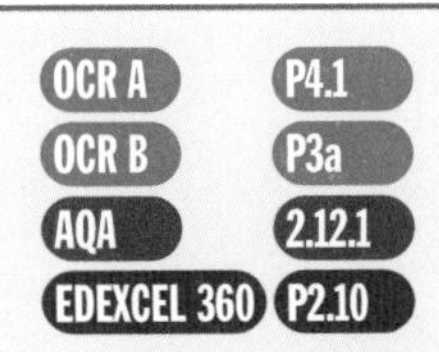

Relativity – how new scientific theories are introduced

Albert Einstein introduced a theory of motion known as the Theory of Relativity. The theory of motion you have studied in this chapter works for everyday speeds and situations. Einstein used his creative imagination to do a thought experiment – it was a new idea not based on any experimental data. For example, he suggested that nothing could travel faster than light – 300 million m/s.

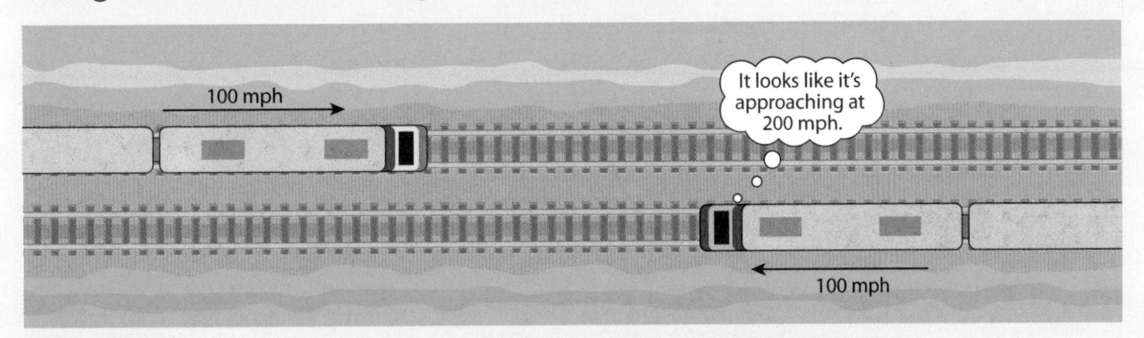

Fig. 6.18 From one train, the other appears to be approaching at 200 mph.

Imagine two trains approaching each other, both travelling at nine tenths (0.9) times the speed of light. If you are travelling on one of them and look at the other, it will appear to be rushing towards you at (0.9 + 0.9 = 1.8) times the speed of light. But, according to Einstein's theory, this is not possible – nothing can appear to travel faster than light. The explanation included mathematical proof that the mass of the trains increased as they got faster, and that time itself passed at a different rate depending on how fast you moved.

Another part of the Relativity Theory says that time passes at a different rate depending on the gravitational field strength. Some scientists were very reluctant to accept the new theories, as, at first, there was no evidence to show these effects. In 1919 the distortion of starlight during an eclipse provided evidence that convinced many scientists.

They also predicted other effects that could be measured to see if the theory was correct. By 1971, an accurate and stable atomic clock was available to test the theory by flying four atomic clocks around the world for three days. When they returned, and the clocks were compared, they confirmed Einstein's Relativity Theory.

This table shows the average of the time differences measured on all the clocks. All the results were within a range of ±23 nm from the average value. The time differences are very small (1 ns = 0.000000001s = 1×10^{-9} s.)

	Average time shift (ns)	
	Travelling East	**Travelling west**
predicted	– 40	+ 275
measured	– 59	+ 273

HOW SCIENCE WORKS

In normal everyday life, you do not notice the effects, but if you use a SatNav system or a GPS receiver they take relativity into account when they work out your position. They need to do this because the signal travels very fast to satellites and back.

1. From the passage on Relativity, write down of an example of something that Einstein said changed, but previous theories said was constant. **[1]**
2. Other scientists were reluctant to accept Einstein's ideas because:
   A they didn't want to change the theory of motion they were using
   B he didn't have any evidence – it was a mathematical theory
   C they each wanted to find a new theory.
   D his evidence didn't match his predictions **[1]**
3. Some scientists said they would not have been convinced if the time measurements had a range of ± 300 nm from the average value. Why not? **[2]**
4. It was over 60 years after Einstein's proposal before these measurements were made to prove the theory. Why was there such a long delay? **[1]**

Exam practice questions

1. This graph shows the movement of a car:

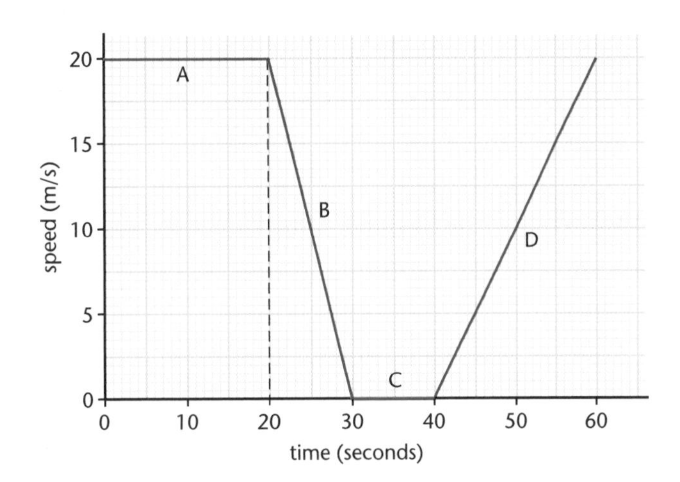

(a) In which section, **A B C** or **D** is the car:

(i) accelerating (speeding up) ?

(ii) stopped? **[2]**

(b) How far does the car travel in the first 20 seconds?

A 0 m **B** 20 m **C** 40 m **D** 400 m **[1]**

(c) The **instantaneous speed** of the car after 50 seconds is:

A 0 m/s **B** 1 m/s **C** 10 m/s **D** 20 m/s **[1]**

2. Choose words from this list to complete the sentences below. A word may be used once, more than once or not at all.

action backward drive equal foot force forward
friction ground opposite pair reaction

Walking involves a ___1___ of forces. The forces are ___2___ and ___3___ and are known as the ___4___ and ___5___ force. The ___6___ pushes ___7___ on the ___8___ and the ___9___ pushes ___10___ on the ___11___. If there is no ___12___ you cannot walk. **[4]**

3. These statements describe what happens during a parachute jump.

A	The parachute opens – increasing the drag force
B	His speed increases and the drag force increases
C	The drag force becomes equal to his weight
D	The skydiver steps from the aircraft and falls, he accelerates at 10 m/s^2
E	He falls with a constant speed
F	His speed decreases

Put them in the correct order. The first has been done for you.

D					

[4]

Exam practice questions

4. Use the equation: momentum = mass × velocity
to match the values of momentum to the situations. The car is travelling in the positive direction:

Situation

A	A car of mass 1000 kg and velocity +28 m/s
B	A lorry of mass 2800 kg and velocity –10 m/s travelling towards a car of mass 1000 kg and velocity +28 m/s
C	A car of mass 1400 kg that is travelling at –10 m/s towards a lorry of mass 2800 kg and velocity + 10 m/s
D	A lorry of mass 2800 kg and velocity +10 m/s travelling towards a car of mass 1000 kg and velocity +10 m/s

Total momentum

1	0 kg m/s
2	28000 kg m/s
3	38000 kg m/s
4	14000 kg m/s

[3]

5. This diagram shows a roller coaster track.

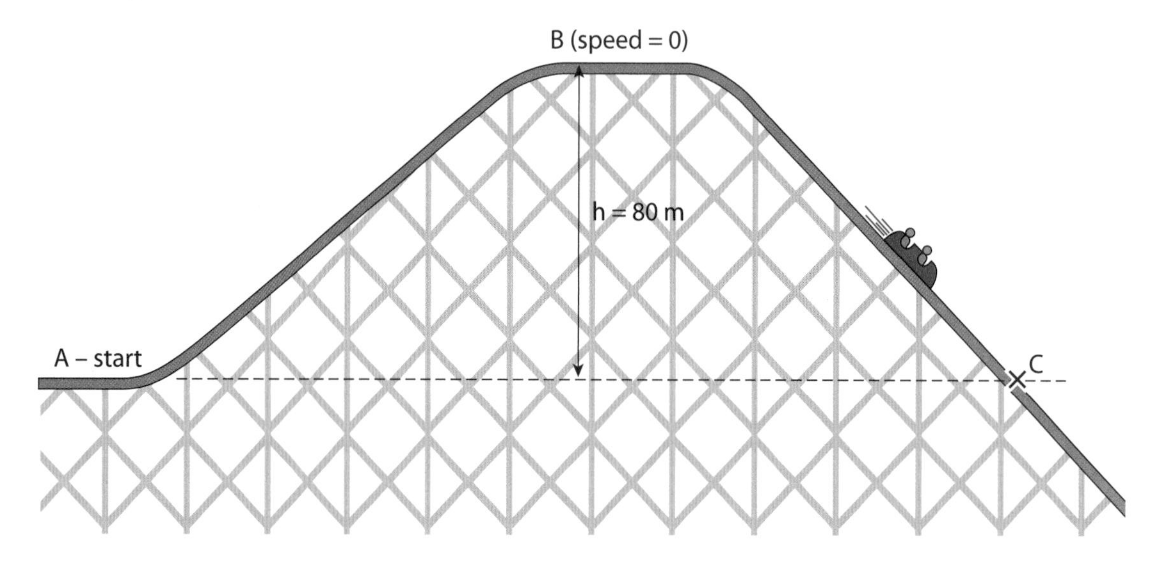

The mass of a car is 800 kg
Gravitational field strength = 10 N/kg

Change in PE = mass × gravitational field strength × height difference
KE = $\frac{1}{2}$ mass × velocity^2
Work done = force × distance

Assume there are no energy losses due to friction as the car goes round the track.

(a) What is the increase in potential energy when the car moves from A to B?
(b) What is the kinetic energy at point C?
(c) What is the speed of the car at point C?
(d) At the end of the ride, the kinetic energy of the car is 25 000 J and it brakes to a stop in 10 m. What is the braking force? **[4]**

Exam practice questions

6. A van with mass 3000 kg travels at a steady speed of 30 m/s.
The resistive force is 5000 N.

(a) What is the driving force on the van in newtons? **[1]**

(b) Using KE = $\frac{1}{2}$ mass × velocity^2 work out the kinetic energy of the van. **[1]**

(c) and **(d)** depend on the specification. Choose either for Edexcel and AQA.

The van stops in 5 seconds.

OCR A **(c)** Using momentum = mass × velocity work out the momentum of the van.

(d) Using force = $\frac{\text{change in momentum}}{\text{time}}$ work out the force stopping the van.

or

OCR B **(c)** Using acceleration = $\frac{\text{change in velocity}}{\text{time}}$ work out the acceleration of the van.

(d) using force = mass × acceleration work out the force stopping the van. **[2]**

7. This table shows the shortest stopping distances for a car at different speeds.

Speed mph	Thinking distance (m)	Braking distance (m)	Stopping distance (m)
20	6	6	12
40	12	24	?
60	18	55	73

(a) Work out the missing value for the stopping distance at 40 mph. **[1]**

(b) These are the shortest stopping distances. Describe a factor that will increase:

(i) the thinking distance

(ii) the braking distance. **[2]**

(c) In a collision test, a dummy with mass 60 kg is travelling at 20 m/s. Work out the kinetic energy of the dummy.
(KE = $\frac{1}{2}$ mass × velocity^2) **[1]**

When the dummy hits the windscreen it stops in 0.001 s
When the dummy is wearing a seatbelt it stops in 0.02 s

(d) What happens to the seatbelt to make the stopping time longer? **[1]**

(e) How does this reduce the damage to the dummy? **[1]**

Studies show that when some drivers wear seat belts, they drive faster and the injuries and deaths of pedestrians increase.

(f) Explain why a driver might increase speed when wearing a seatbelt. **[2]**

Hannah travels every day by car. She does not want to travel by train because she thinks it may crash. This chart gives information about deaths on the roads and railways in the UK in 2004.

UK 2005	Roads	Railways
Deaths	1675	5
Death rate	3.3 travellers per billion vehicle kilometres	0.1 travellers per billion passenger kilometres

(g) In the table, which has higher risk, road or rail travel? **[1]**

(h) Suggest a reason why Hannah thinks that rail travel is more dangerous than road travel. **[2]**

Chapter

7 Electricity

The following topics are covered in this chapter:

- *Electrostatics*
- *The mains electricity supply*
- *Electric circuits*

7.1 Electrostatics

Electric charge

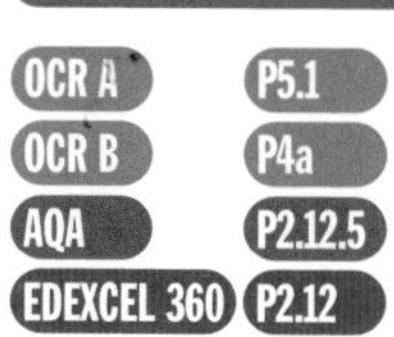

Electric charge can be **positive** or **negative**. **Electrons** are particles with a negative electric charge. They can move freely through a **conductor**, for example, any type of **metal**, but no charged particles can move through an **insulator**.

Materials that are positively charged have missing electrons, materials that are negatively charged have extra electrons

Electrostatic effects are caused by the transfer of **electrons**. (This is sometimes called **static electricity**.) When **insulators** are rubbed, **electrons** are rubbed off one material and **transferred** to the other.

Fig. 7.1 shows a polythene rod rubbed with a duster. It has picked up **electrons** from the duster and becomes negatively charged leaving the duster positively charged.

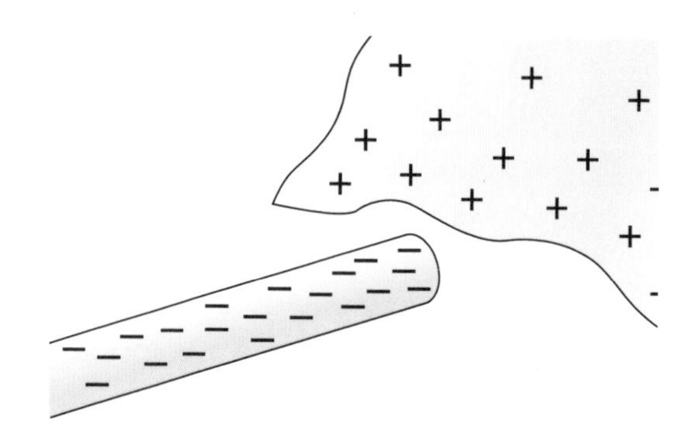

Fig. 7.1 The insulated rod and the cloth have opposite charge.

Conductors cannot be charged unless they are completely surrounded by **insulating materials**, like dry air and plastic – otherwise the electrons flow to or from the conductor to discharge it. An **insulated conductor** can be charged by rubbing with a **charged duster**, or touching it with a **charged rod**. Some electrons will be transferred, so that the charge is spread out over both objects.

Two objects attract each other if one is positively charged and the other is negatively charged. Two objects with similar charge (both positive or both negative) repel.

Remember: Like charges repel and unlike charges attract.

A conductor can be **discharged** by touching it with another conductor, such as a wire, so that electrons flow along the wire and cancel out the charge.

The earth connection

To stop conductors becoming charged, we sometimes use a thick metal wire to connect them to a large metal plate in the ground. This acts as a large reservoir of electrons. We say the object is connected to **earth**, or **earthed**.

Electrons flow so quickly to or from earth that earthed objects do not become charged. If all the metal water pipes in a house are connected into the ground like this, they can be used as an earth.

Dangers of electrostatic charge

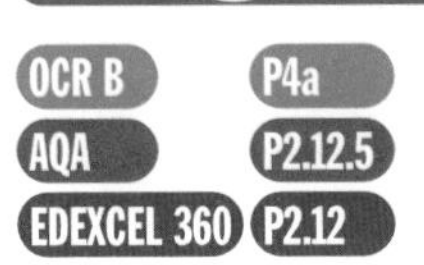

The **human body** conducts electricity. When the flow of charge is large enough for our **nerves** and **muscles** to be affected, we call this an **electric shock**. **Electrostatic shocks** are usually small and not harmful. For example those from touching a car door on a dry day when charge has built up on the metal car and flows through you to earth. Larger ones can be dangerous to people with heart problems because a flow of charge through the body can **stop** the **heart**. **Lightning** is a very large **electrostatic discharge**. When it flows through a body it is often fatal.

If you are standing on an **insulating mat** or wearing shoes with **insulating soles** when you touch a charged object, then this will reduce the chance of an electric shock because the charge will not flow through you to earth. You will become charged, and stay charged until you touch a conductor.

Explosions

When a charged object is close to a conductor, electrons can jump across the gap. This is a spark and can cause an explosion if there are:

inflammable vapours like petrol or methanol

powders in the air, like flour or custard, which contain lots of oxygen – as a dust they can explode

inflammable gases like hydrogen or methane

Fig. 7.2 sparks can cause explosions.

Lorries containing inflammable gases, liquids and powders are connected to earth before loading or unloading. Aircraft are earthed before being refuelled. This prevents charge from building up on metal pipes or tanks when the loads are moved, so there is no danger of a spark igniting the load.

Annoying electrostatic charge

Charged objects attract small particles of dust and dirt, for example, plastic cases and TV monitors.

Clothing can be charged as you move and 'clings' to other items of clothing, or the body. Synthetic fibres are affected more than natural fibres as they are better insulators.

Anti-static sprays, liquids and cloths stop the build up of static charge. These work by increasing the amount of conduction – sometimes by attracting moisture because water conducts electricity.

Uses – photocopiers and laser printers

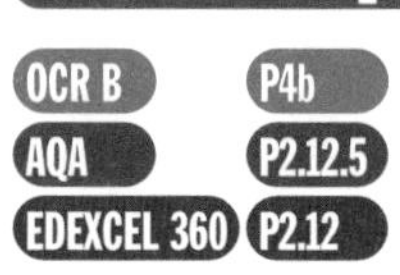

In a **photocopier**:

- a rubber belt is coated with a material that conducts electricity only when it is illuminated
- the belt is charged
- a bright light is used to make an image of the sheet of paper to be copied on the belt

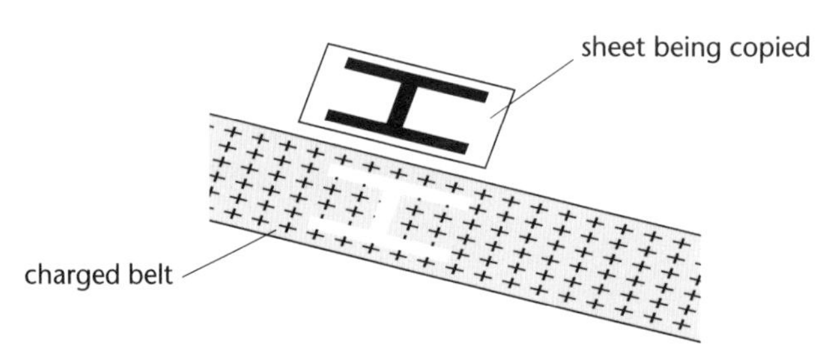

Fig. 7.3 Making a photocopy.

- the illuminated parts of the belt discharge
- the belt is sprayed with black powder that sticks to the charged areas
- a sheet of paper is pressed on the belt so the black powder is transferred
- the paper is heated to make the powder stick to the sheet.

A **laser printer** is very similar, but instead of a bright light making an image of the page, a laser light beam is used to write the characters on the belt.

Uses – Electrostatic precipitators

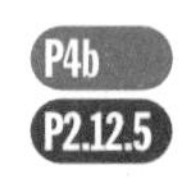

Electrostatic precipitators remove dust or smoke particles from chimneys, so that they are not carried out of the chimney by the hot air.

- Metal plates or grids are put in the chimneys.
- They are charged by connecting them to a high voltage.
- The smoke particles are attracted to the charged plates or grids.
- The particles clump together on the plates to form larger particles.
- When they are heavy enough, the smoke particles fall back down the chimney into containers.

The grids are positively charged in some designs and negatively charged in others.

Uses – Paint spraying

OCR B P4b

The **paint** and the object are given different charges so that the paint is **attracted** to the object.

- The spray gun is charged so that it charges the paint particles.
- The paint particles repel each other to give a fine spray.
- The object is charged with the opposite charge to the paint.
- The object attracts the paint.
- The paint makes an even coat, it even gets underneath and into parts that are in shadow.
- Less paint is wasted.

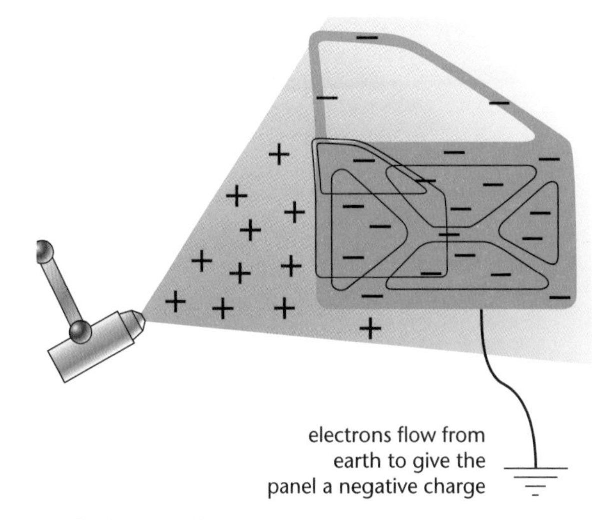

Fig. 7.4 Paint spraying.

Uses – Defibrillators

OCR B P4b

When the heart beats, it is the heart muscle contracting. A **defibrillator** is used to start the heart when it has stopped.

- Two electrodes called paddles are placed on the patient's chest.
- Everyone, including the operator must 'stand clear' so they don't get an electric shock.
- The paddles are charged.
- The charge is passed from one paddle, through the chest to the other paddle to make the heart muscle contract.

The paddles must make a good electrical contact with the patient's chest.

Uses – Fingerprinting

EDEXCEL 360 P2.12

If a dust print is left on a surface this can be lifted using charged **lifting film**.

- The lifting film is made of material that stores electrostatic charge.
- The film is placed over the print.
- The dust particles are attracted and stick to the film.
- The print is stored on the film.

7.2 Electric circuits

Circuit symbols

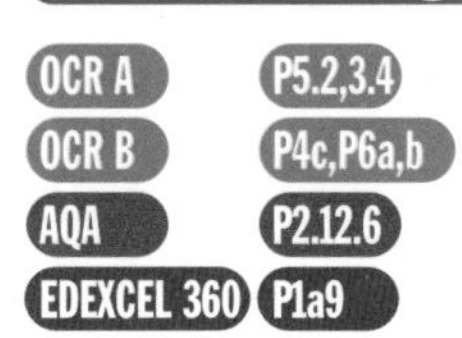

More than one cell used to be called a 'battery of cells'. Now 'battery' is often used to mean one cell. OCR A refers to one battery, but other specifications expect you to use the word 'cell'.

For OCR A you need to know that current is a flow of positive charge so the direction of the current is opposite to the direction of the electron flow, because electrons are negatively charged.

Component	Symbol	Component	Symbol
switch (open)		lamp	
switch (closed)		fuse	
cell		fixed resistor	
battery		variable resistor	
ammeter	A	light dependent resistor (LDR)	
voltmeter	V	thermistor	
junction of conductors		diode	
motor	M	generator	G
power supply		a.c. power supply	

Electric current and charge

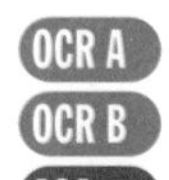
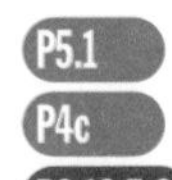

Electric current:

- is a flow of electric **charge**
- only flows if there is a **compete circuit**. Any break in the circuit switches it off
- is measured in **amperes** (A) using an **ammeter**
- is not used up in a circuit. If there is only one route around a circuit, the current will be the **same** wherever it is measured
- transfers energy to the components in the circuit.

Some people prefer not to say 'current flows' because this means 'a flow of charge flows.' Others find it helps them to picture what is happening.

An electric current is a flow of electric charge.

Series circuits

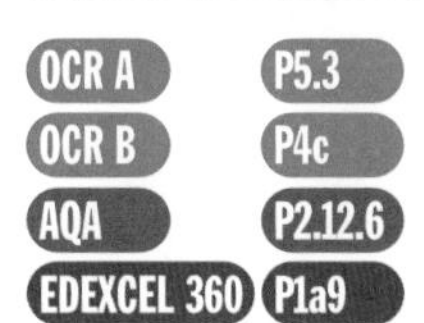

A **series** circuit is a circuit with only one route around it. Fig. 7.5 shows a lamp and motor connected in series to a battery with two ammeters in series to measure the current through the lamps. The current measured on each ammeter will be the same.

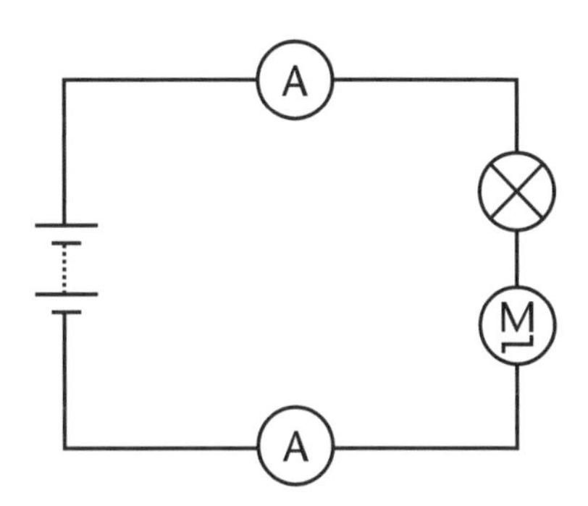

Fig. 7.5 A series circuit.

Parallel circuits

AQA P2.12.6

A **parallel** circuit has more than one path for the current around the circuit. Fig. 7.6 shows a motor and a lamp connected in parallel to a battery. There are two paths (marked in red and blue) around the circuit so the current measured on ammeters B and C adds up to give current measured on ammeter A and on ammeter D.

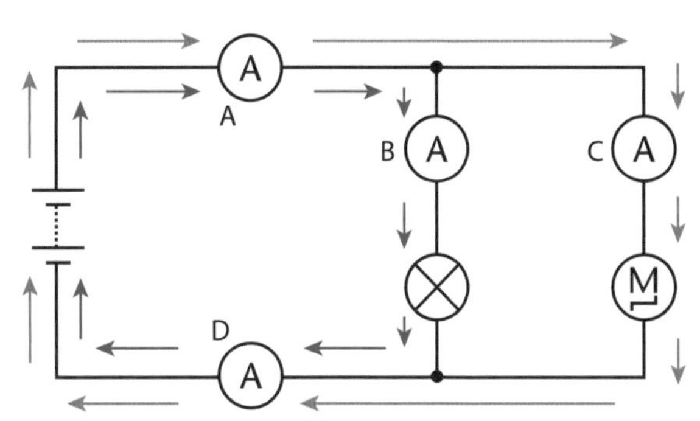

Fig. 7.6 A parallel circuit.

Voltage (also called potential difference)

OCR A P5.3
AQA P2.12.5,6
EDEXCEL 360 P3.6

Voltage is also called **potential difference (p.d)**. It is:

- measured between two points in a circuit
- measured in **volts** (V) using a **voltmeter**
- a measure of **energy transferred** to (or from) the charge moving between the two points
- measured between the terminals of a battery, or other power supply, when the energy is transferred to the charge
- measured between the ends of a component when the energy is transferred from the charge.

The higher the voltage of a battery, the higher the 'push' on the charges in the circuit.

When components are connected in **series**, as shown in Fig. 7.7, the voltage, or potential difference (p.d.) of the power supply is shared between the components.

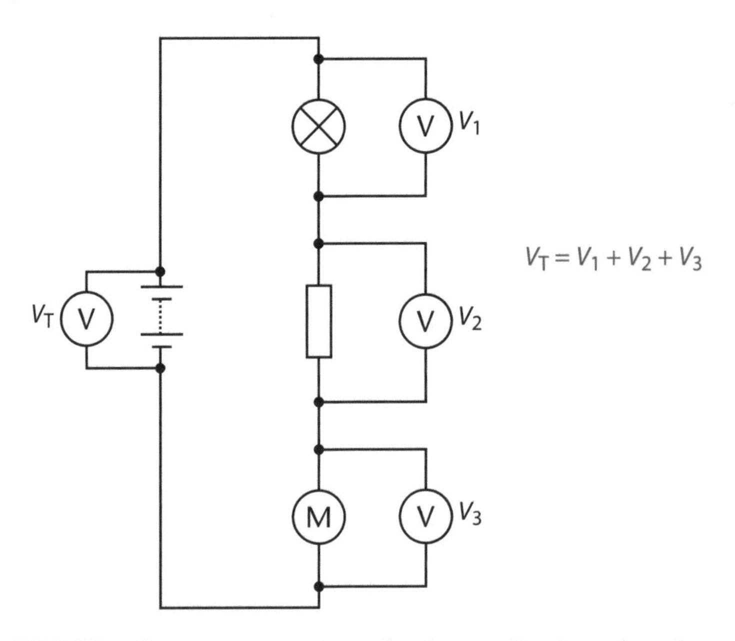

Fig. 7.7 Adding the measurements on the three voltmeters gives the power supply p.d.

Remember, it is 'current through' but 'voltage across' a component.

When components are connected in parallel to a power supply, as shown in Fig. 7.8, the voltage, or potential difference (p.d), across each component is the same as that of the power supply.

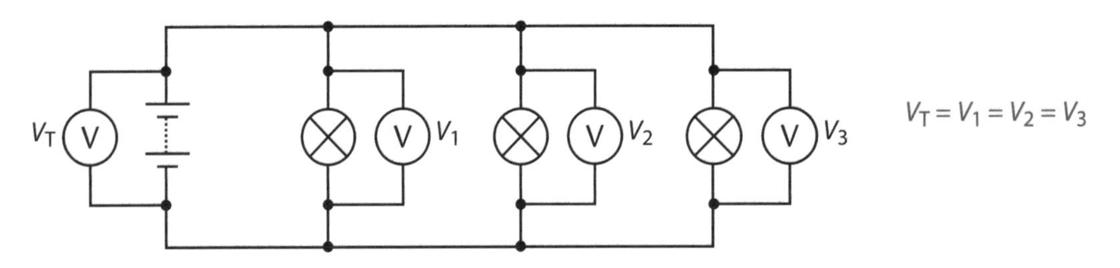

Fig 7.8 The measurements on all the voltmeters are the same.

Ammeters are connected in series. Voltmeters are connected in parallel.

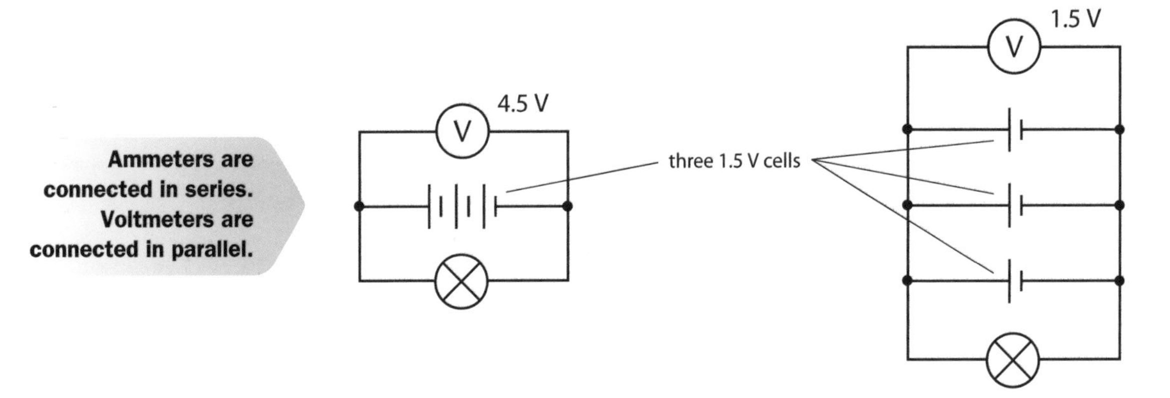

Fig. 7.9 Cells connected in series and in parallel.

Resistance and resistors

OCR A P5.2
OCR B P4c,P6a
AQA P2.12.6
EDEXCEL 360 P1a9

All the components and wires in a circuit **resist** the flow of electric charge through them.

When the **voltage** is fixed (for example, by using a battery) the larger the **resistance** of a circuit, the less **current** passes through it.

The resistance of the connecting wires is so small it can usually be ignored. Other metals have a larger resistance, for example the filament of a light bulb is very large. **Metals** get **hot** when charge flows through them. The larger the resistance, the hotter they get. A light bulb filament gets so hot that it glows.

KEY POINT

$$\text{resistance} = \frac{\text{voltage}}{\text{current}}$$

Resistance is measured in ohms (Ω) where current is in amperes (A) and voltage in volts (V)

$R = \frac{V}{I}$ **and also** $V = IR$ **and** $I = \frac{V}{R}$

For how to rearrange the formula see page 225.

Fixed resistors

OCR A P5.2
OCR B P6a
AQA P2.12.6
EDEXCEL 360 P1a9

In some components, such as **resistors** and **metal conductors**, the resistance stays constant when the current and voltage change, providing that the temperature does not change.

For this type of fixed resistance if the **voltage** is increased (for example, by adding another battery), the **current** increases. A graph of current against voltage is a straight line, because the **current** is **directly proportional** to the **voltage** – doubling the voltage doubles the current. Components that obey this law (sometimes called Ohm's Law) are sometimes called **ohmic** components or devices.

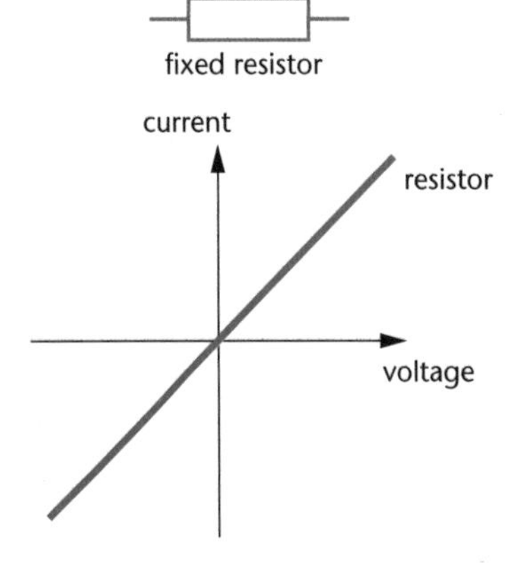

Fig. 7.10 A graph of current against voltage for a resistor.

Changing resistance

OCR A P5.2
OCR B P4c,P6a
AQA P2.12.6
EDEXCEL 360 P1a9

Variable resistors

A **variable resistor** changes the current in a circuit by changing the resistance. This can be used to change how circuits work, for example, to change:

- how long the shutter is open on a digital camera
- the loudness of the sound from a radio loud speaker
- the brightness of a bulb
- the speed of a motor.

Inside one type of variable resistor is a long piece of wire made of metal with a large resistance (called **resistance wire**). To alter the resistance of the circuit a sliding contact is moved along the wire to alter the length of wire in the circuit, as shown in Fig. 7.11.

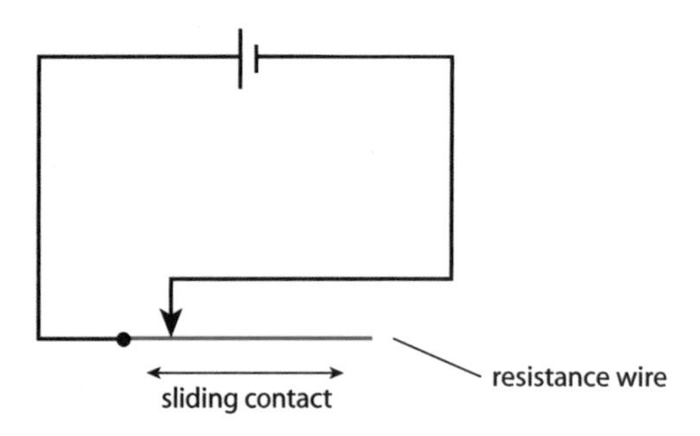

Fig. 7.11 A variable resistor design.

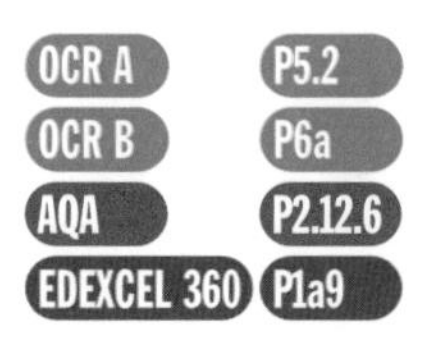

Filament lamps

The wire in a **filament lamp** gets hotter for larger currents. This increases the resistance so the graph of current against voltage is not a straight line.

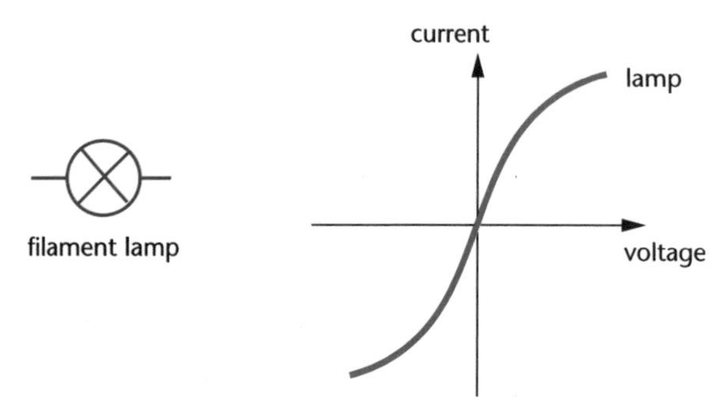

Fig. 7.12 A graph of current against voltage for a filament lamp.

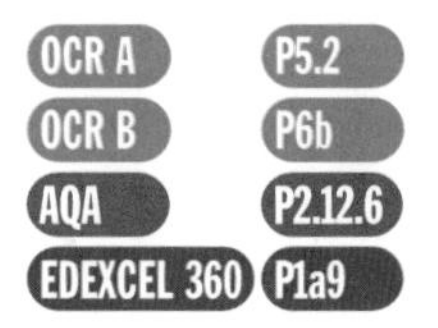

Light dependent resistor (LDRs)

The resistance of a **light dependent resistor (LDR)** decreases as the light falling on it increases. This can be used in a circuit to control when a lamp switches on or off.

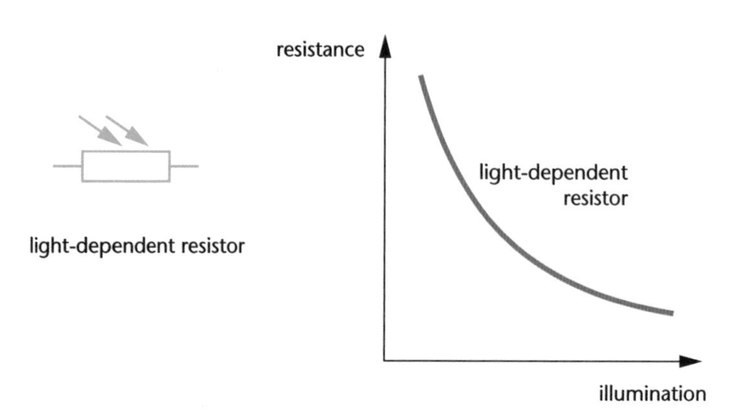

Fig. 7.13 A graph of resistance against intensity of light for a light-dependent resistor.

Thermistors

OCR A P5.2
OCR B P6b
AQA P2.12.6
EDEXCEL 360 P1a9

The resistance of a negative temperature coefficient (NTC) **thermistor** decreases as the temperature increases. This can be used to switch on a heating or cooling circuit at a certain temperature.

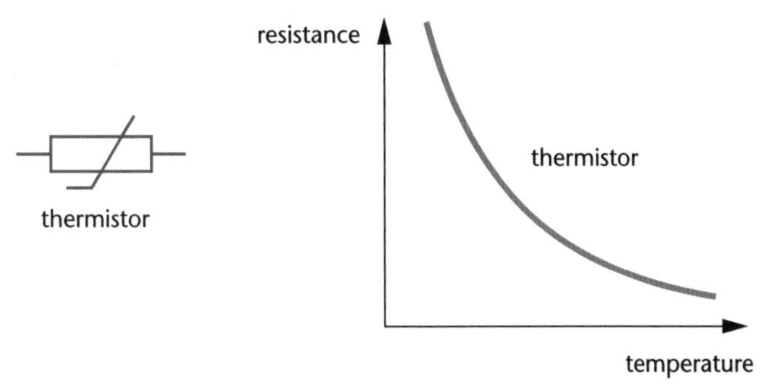

Fig. 7.14 A graph of resistance against temperature for a thermistor.

Diode

Current will only flow through a **diode** in one direction. In one direction its resistance is very low, but in the other direction, called the **reverse direction**, its resistance is very high.

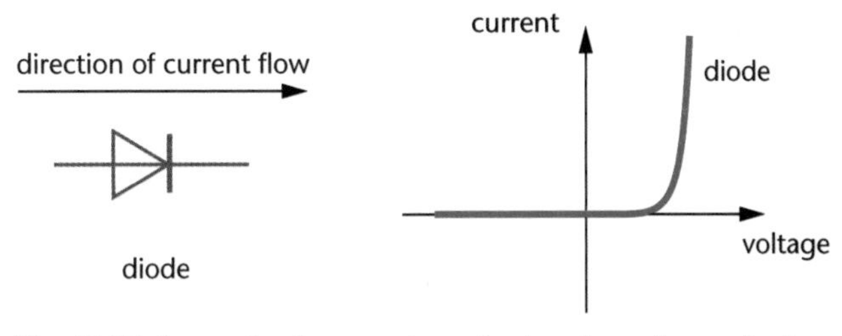

Fig. 7.15 A graph of current against voltage for a diode.

Combining resistors

Components can be added to a circuit in series or in parallel, as shown in Fig. 7.16.

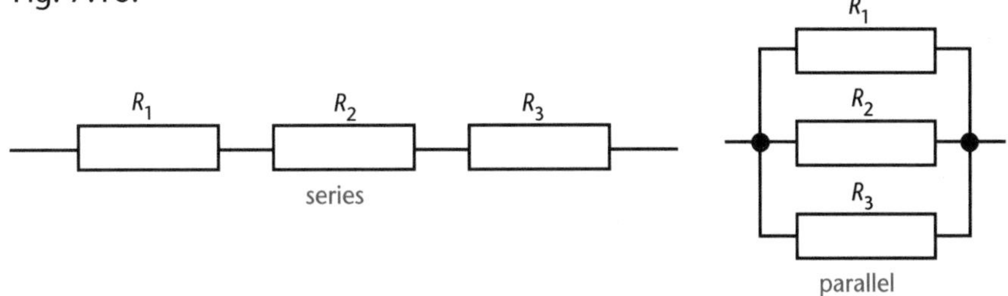

Fig. 7.16 Resistors connected in series and in parallel.

For components in **series**:

- Two (or more) components in **series** have more resistance than one on its own because the battery has to push charges through both of them.
- The **current** is the **same** through each component.
- The **p.d.** is **largest** across the component with the **largest resistance** because more energy is transferred by the charge passing through a large resistance than through a small one.

- The **p.d.s** across the components **add** up to give the p.d. of the **power supply**.

When components are connected in series, the total resistance of the circuit is equal to sum of the resistances of all the components.

$R_T = R_1 + R_2 + R_3$

Where R_T is the total resistance of components with resistances R_1, R_2 and R_3 connected in series.

For components in **parallel**:

- A combination of two (or more) components in **parallel** has less resistance than one component on its own. This is because there is more than one path for charges to flow through.
- The **current** through each component is the same as if it were the only component present.
- The **total current** will be **sum** of the currents through all the components.
- The **p.d.** across all the components will be the **same** as the power supply **p.d.**
- The **current** is **largest** through the component with the **smallest resistance**. This is because the same battery voltage makes a larger current flow through a small resistance than through a large one.

A theory of resistance

Metals are made of a fixed **lattice** of atom centres surrounded by **free electrons**. The moving electrons form the current. In metals with low resistance, the electrons require less of a 'push' (p.d) to get through the lattice. The moving electrons **collide** with the stationary atom centres and make them **vibrate more**. This is a heating effect and increases the temperature of the metal.

7.3 The mains electricity supply

Safe use of mains electricity

The table shows the colour code for the mains electricity cables used in buildings, and in appliances.

Name of wire	Colour of insulation	Function of the wire
live	brown	carries the high voltage
neutral	blue	the second wire to complete the circuit
earth	green and yellow	a safety wire to stop the appliance becoming live

Fig. 7.17 shows how the plug is wired for a heater with a metal case. The **fuse** is always connected to the brown, **live** wire. A **cable grip** is tightened where the cable enters the plug to stop the wires being pulled out.

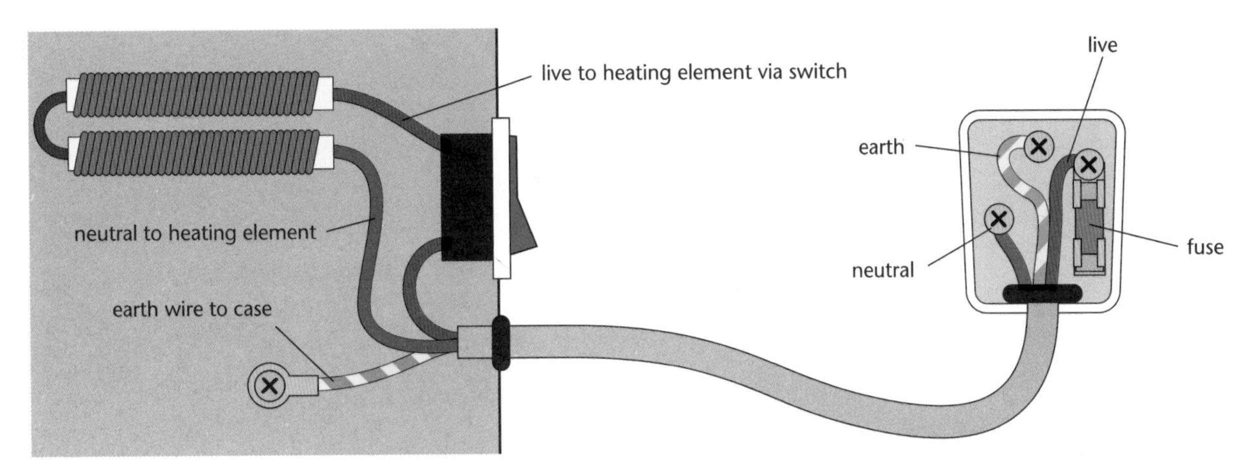

Fig. 7.17 A 3-pin plug on an earthed metal appliance.

KEY POINT

A fuse is a piece of wire that is thinner than the other wires in the circuit and will melt first if too much current flows and the wires overheat. A 3A fuse will melt if a current of 3A flows through it. The value of the fuse should be the lowest value that is more than the normal operating current.

For AQA you need to decide the correct fuse. Calculate the current and choose the lowest fuse that is higher than this value.

If there is a fault, or if too many appliances are plugged into one socket, resulting in a large current, then the fuse will melt and break the circuit preventing a fire.

The **earth** wire is connected to the metal case of metal appliances so that when they are plugged into the mains supply, the metal case is earthed (see 7.1 The earth connection) If there is a fault and the live wire touches the metal case, a very large current flows through the low-resistance path to earth melting the fuse wire and breaking the circuit.

Double insulated appliances have cases that do not conduct (usually plastic) and no metal parts you can touch so they do not need an earth wire.

The **fuse** takes a short time to melt, so it will not prevent you from getting an electric shock, which can be fatal, if you touch a live appliance.

Appliances that are particularly dangerous are those like lawn mowers and power tools, where the cable could get wet, or be cut. Music amplifiers are also dangerous where you are touching the metal instrument and there may be a lot of electrical equipment around that has not been safety tested. These should be connected using an **RCD (residual current device)** also called a **RCCB (residual current circuit breaker)**. These are switches to cut off the electricity very quickly if they detect a difference in the current flowing in the live and the neutral wires. (This would happen, for example, if the current was flowing through a person, or appliance casing.) Another advantage is that they can be switched back on once the fault is fixed, whereas a fuse must be replaced. An **RCD** can be part of a mains circuit in a building, or a plug-in device that goes between the appliance and the socket.

Power, current and voltage

OCR A P5.5
AQA P2.12.8

For how to rearrange the formula see page 225.

KEY POINT

Electrical power is worked out using:

Power = current × voltage

P = IV

Power is measured in watts (W), current in amperes (A) and voltage in volts (V).

Example
What is the current in a 2.8 kW kettle?

Using $P = I\,V$

$I = P \div V$

$I = 2800\,\text{W} \div 230\,\text{V} = 12.2\,\text{A}$

For OCR A you need to revise efficiency, from section 2.3, so you can calculate efficiency of electrical appliances.

Example
A 60 W light bulb is 5% efficient. How many watts of light does it emit?

$$\text{Efficiency \%} = \frac{\text{useful energy output}}{\text{total energy input}} \times 100$$

1 watt = 1 joule per second, so in one second:

$$\text{useful energy output} = \frac{5 \times 60\text{J}}{100} = 3\text{J}$$

Energy and charge

AQA P2.12.8

KEY POINT

Electric charge is measured in coulombs (C)

charge (C) = current (A) x time(s)

KEY POINT

The energy transformed in a circuit depends on the charge that has flowed and the potential difference.

energy transformed (J) = potential difference (V) x charge (C)

More about electromagnetic induction

OCR A P5.4
OCR B P6d
AQA P3.13.8

Electromagnetic induction is the name of the process where a changing magnetic field induces a voltage in a conductor. It is used in generators and in transformers. If there is an **induced voltage** in a coil and the ends are connected to make a complete circuit then a current flows in the coil.

This section builds on section 2.4 How generators work and a.c. and d.c.

Fig. 7.18 shows how the current in the coil of a generator changes with the rotation of the magnet or electromagnet.

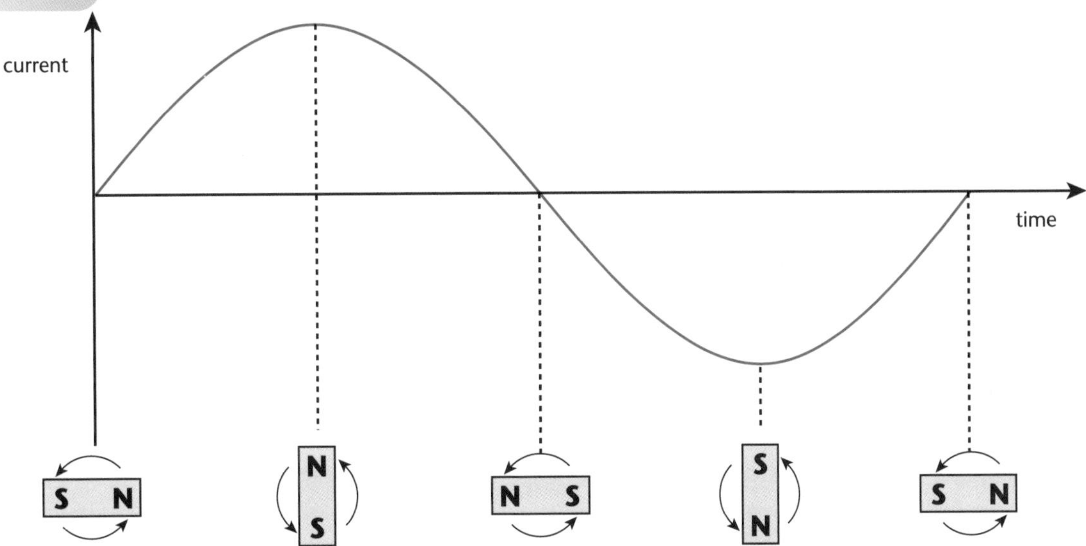

Fig. 7.18 How a rotating magnet changes the current in the coil.

In a generator, the induced voltage can be increased by:

- increasing the speed of rotation of the magnet or electromagnet
- increasing the strength of the magnetic field
- increasing the number of turns on the coil
- placing an iron core in the coil.

The a.c. mains supply voltage is 230 V. We use a.c. for the mains supply because it is easier to generate than d.c. and can be distributed more efficiently.

Transformers

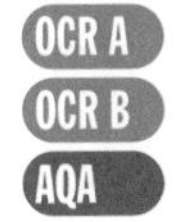

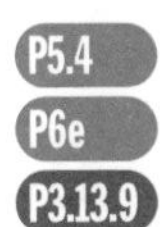

A changing magnetic field is needed so transformers will not work with a d.c. supply.

There are three types of transformer. **Step-up transformers**, which increase voltage, and **step-down transformers**, which decrease voltage, are covered in section 2.4. The third type of transformer is called an **isolating transformer**. It does not change the voltage, but is used for safety reasons. The transformers in bathroom shaver sockets are isolating transformers so that there is less risk of contact between the live wire and the Earth wire – for example if the plug got wet.

Fig. 7.19 shows how a transformer is made. There are two coils, **a primary coil** and **a secondary coil**. The iron core concentrates the magnetic field in the coils.

- An alternating current in the primary coil produces a changing magnetic field.
- The changing magnetic field induces a voltage in the secondary coil.

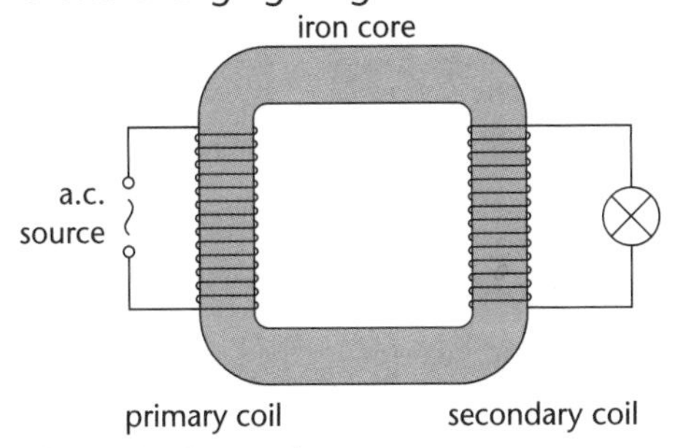

Fig. 7.19 A transformer.

KEY POINT

The ratio of the voltages across the primary and secondary coils is the same as the ratio of the number of turns in the coils.

$$\frac{\text{primary voltage}}{\text{secondary voltage}} = \frac{\text{number of turns in primary coil}}{\text{number of turns in secondary coil}}$$

$$\frac{V_P}{V_S} = \frac{N_P}{N_S}$$

Step-up transformers have **more** turns in the **secondary** coil, **step-down** transformers have more turns in the **primary** coil, and **isolating** transformers have the **same** number of turns in both coils.

The induced voltage in the secondary coil is an alternating voltage with the same frequency as the voltage in the primary coil.

Power transmission

 OCR B  P6e

Energy is conserved, so every second, if there are no losses to heat and sound, all the energy is transferred from the primary coil to the secondary coil.

KEY POINT

The power in the secondary coil is equal to the power in the primary coil.
$V_pI_p = V_sI_s$
Where V_p and I_p are the voltage and current in the primary coil, and V_s and I_s are the voltage and current in the secondary coil

(In fact, large transformers get hot, and hum, but transformers are very efficient compared to other stages of electricity generation, so you can assume 100% efficiency unless you are told otherwise.)

This relationship tells us that if the voltage is stepped-down the current is increased, if the voltage is stepped-up the current is decreased.

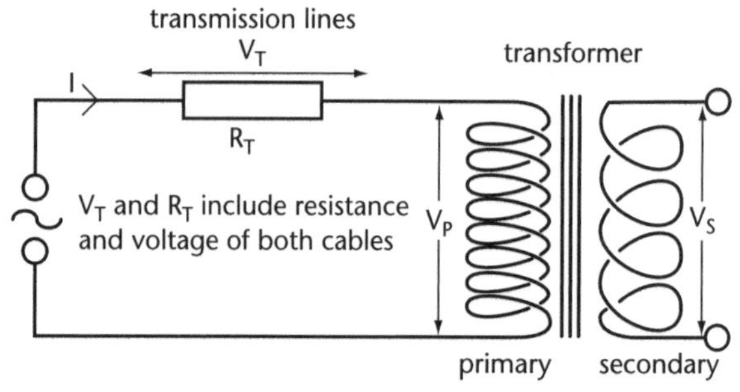

Fig 7.20 The current and voltages in a transmission line and transformer

Fig. 7.20 shows how this is made use of in transmission lines to reduce power loss. By increasing the primary voltage V_p the current in both the primary coil and the transmission lines is made very small. The power lost in the transmission lines, P_T is

$$P_T = I\,V_T$$

Where V_T is the voltage across the transmission lines $V_T = I\,R_T$
(R_T is the total resistance of the transmission lines)

$$P_T = I \times (I\,R_T)$$
$$P_T = I^2\,R_T$$

KEY POINT

The power lost in the transmission lines depends on the current squared, so by making the current as small as possible the power loss is reduced.

HOW SCIENCE WORKS

OCR A P5.2
OCR B P4c
AQA P1a9
EDEXCEL 360 P1a9

Jack and Emily set up a circuit to measure the current through a resistor for different values of the voltage across the resistor. Between each measurement, they switch off the circuit to make sure that it doesn't heat up. Fig. 7.21 shows the circuit they used, their results and a graph they plotted.

Voltage (volts)	Current (milliamps)			
	1st reading	2nd reading	3rd reading	Average reading
0	0	0	0	0
0.5	2.0	2.6	2.0	2.2
1.0	5.0	5.2	4.8	5.0
1.5	7.5	7.4	7.6	7.5
2.0	10.3	8.8	9.9	10.1
2.5	11.8	12.4	13.0	12.4
3.0	14.8	15.0	15.5	15.1

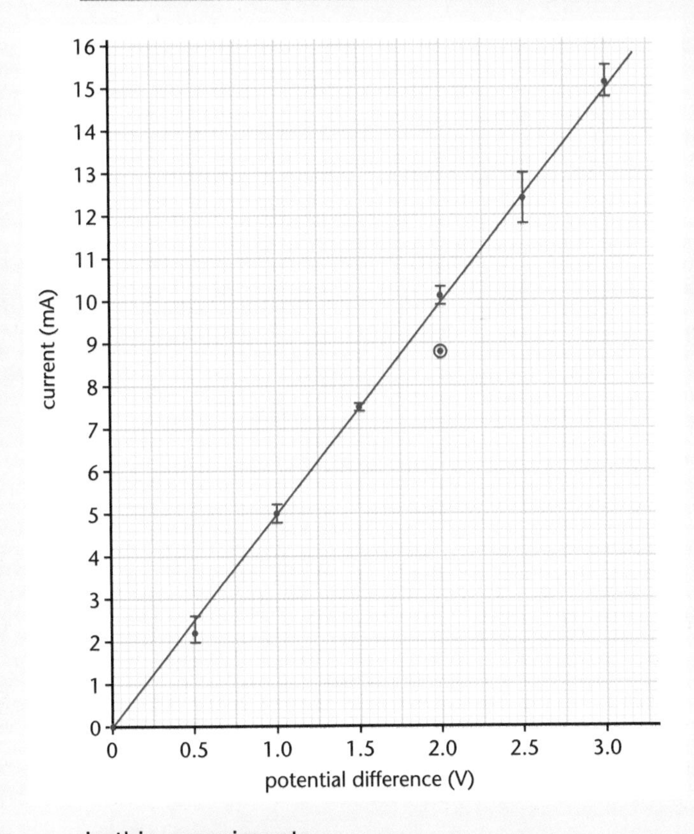

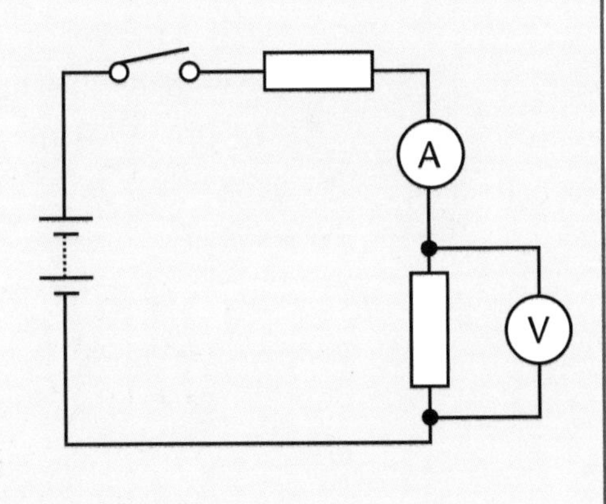

Fig. 7.21

In this experiment:

- voltage is the **independent variable** (the variable that Jack and Emily changed), so it is plotted on the horizontal axis
- current is the **dependent variable** (it changes when the voltage changes), so it is plotted on the vertical axis
- temperature is a **control variable** (it will affect the results if it changes, so Jack and Emily take care to keep it constant)
- all of these **variables** are **continuous** (or **contiguous**), they can take any value on a sliding scale. A **discrete variable** is one that has only set values, for example the number of components in the circuit, or the number of measurements taken.

HOW SCIENCE WORKS

To calculate an **average** measurement for the current, Jack adds the three measurements together and divides by three, because there are three measurements.

He says that one of the values for the current is obviously wrong, so he is not including it. A result like this is called an **outlier** or an **anomalous result**. He calculates an average of the other two measurements.

Emily plots error bars on the points to show the **range** of the measurements. The smaller the range, the more **reliable** the average measurement. She draws a **best straight line** through the points. This is one straight line, not a lot of separate straight sections.

1. Which of these variables is discrete?
   A number of batteries
   B voltage
   C current
   D electrical resistance **[1]**
2. Which of the measurements for current and p.d. has Jack decided is an outlier? **[1]**
3. What is the p.d. for the current measurement that has the greatest range? **[1]**
4. Emily takes one more measurement at 3.5 V. The current measurements are 17.3 mA, 17.9 mA and 18.2 mA. Work out **(a)** the range and **(b)** average current value. **[2]**

Exam practice questions

1.
A
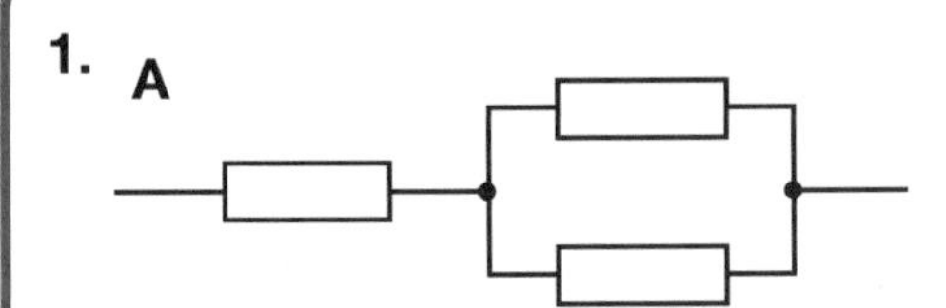
B
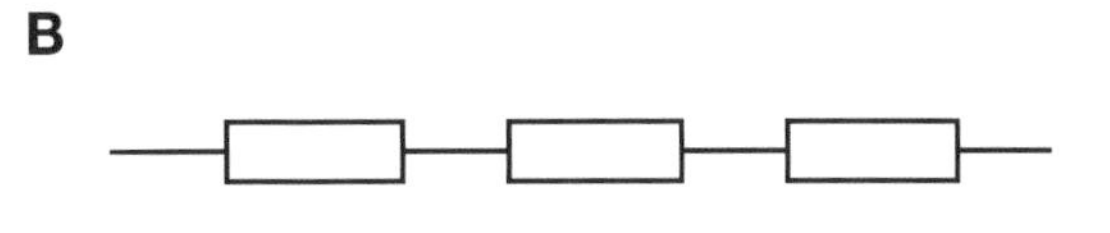
C
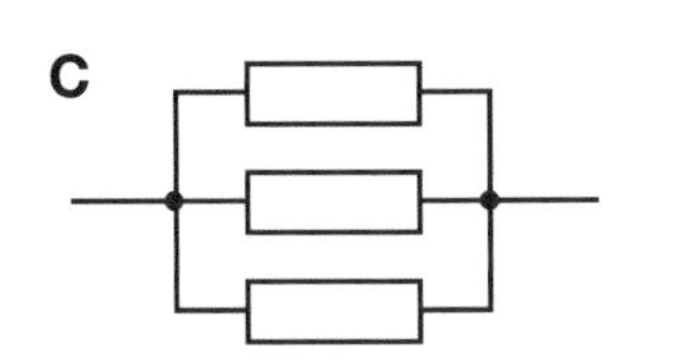
D
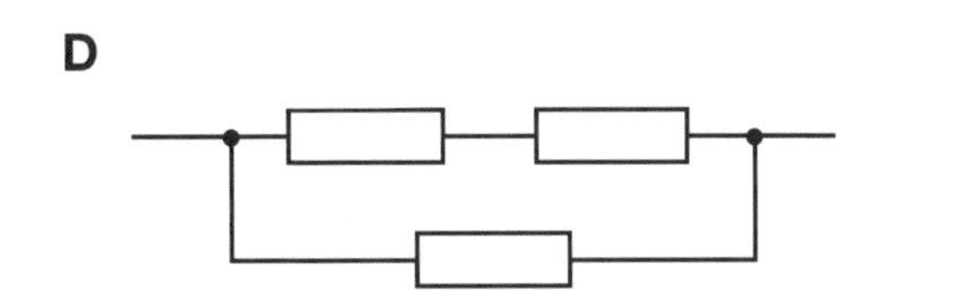

All the resistors in these combinations are identical. Which of the combinations has the largest resistance? [1]

2. The energy wasted in electrical cables can be reduced by transmitting energy through cables with:

A a high current
B a high heat capacity
C a high resistance
D a high voltage [1]

3. What size fuse should be fitted to a plug for a 500 W hairdryer?
(mains supply voltage = 230 V)

A 1 A
B 3 A
C 5 A
D 13 A [1]

4. Draw straight lines to link the **electrical component** to the correct **symbol**

Electrical component	**Symbol**
light dependent resistor (LDR)	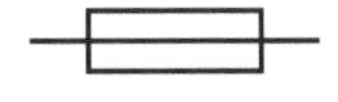
thermistor	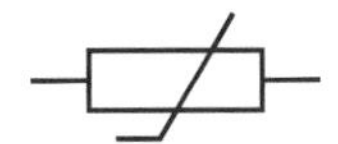
fuse	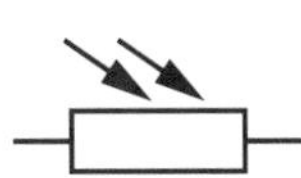

[3]

5. Put these steps in order to describe how a transformer works.

A The induced voltage makes an alternating current flow in the coil.
B The alternating voltage across the primary coil produces a changing magnetic field.
C The changing magnetic field induces an alternating voltage in the secondary coil.
D The magnetic field changes in the secondary coil.
E The magnetic field changes in the iron core. [4]

Exam practice questions

6. Choose words from the list to complete this description of how a defibrillator can restart the heart. (Use words once, more than once or not at all.)

charge **chest** **contract** **electrical** **head** **shock** **stop** **voltage**

Paddles are placed on the patient's ____1____ . They must make a good ____2____ contact with the body. To make the heart ____3____ electric ____4____ is passed through the body. No-one must touch the person so they do not get an electric ____5____ . **[5]**

7. Sadia has bought a new electric lamp. It is **double insulated** so it does not need **earthing**.

(a) What does double insulated mean? **[1]**

(b) The lamp has a 3 A fuse. Describe what happens when:

(i) The normal current of 0.25 A flows in the circuit

(ii) There is a fault and the current increases to 5 A. **[2]**

(c) The fault is fixed and the fuse is replaced with a 13 A fuse. Explain why this is not a good idea. **[1]**

8. Tom works at a car body shop. He spray paints the cars.
The spray gun is charged so that the **paint droplets** are all **positively charged**.

(a) What effect does this have on the paint spray? **[2]**

(b) The car body is charged with the **opposite charge** to the paint droplets. Why is this? **[1]**

(c) Describe two advantages of charging the paint spray and the car body **[2]**

9. Ali is investigating how the current through a diode depends on the potential difference.

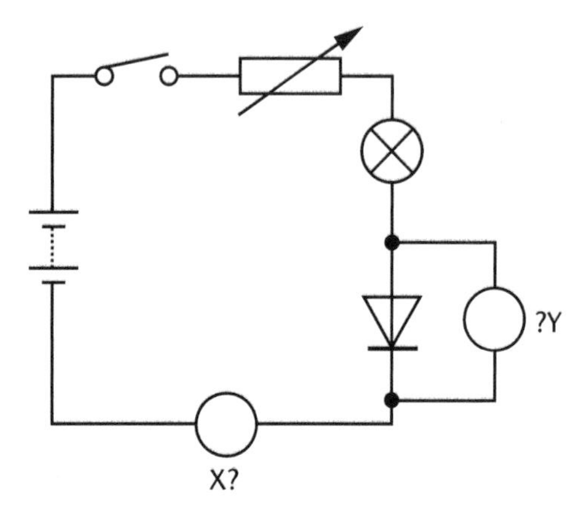

(a) Complete the circuit to show where the ammeter and the voltmeter should be placed in the circuit. **[2]**

Exam practice questions

He plots this graph for the diode to show how the current varies with potential difference.

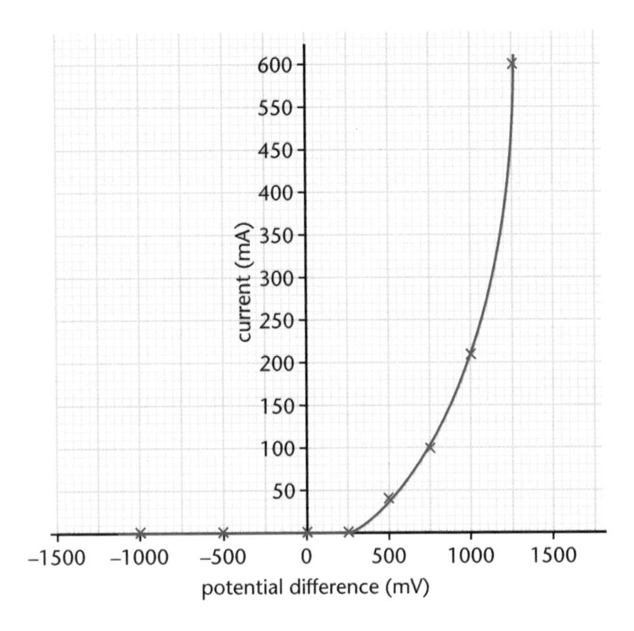

The lamp has a normal operating current of 250 mA. At this current it has a normal brightness.

(b) What is the potential difference that gives a current of 250 mA? **[1]**

(c) Describe how the brightness is changed when the potential difference is

(i) 1200 mV

(ii) – 1200 mV

(iii) 500 mV **[3]**

(d) Ali replaces the battery with an a.c. power supply. How is an a.c. supply different to a d.c. supply? **[2]**

(e) Explain what will happen to the current now that there is an a.c. supply and a diode in the circuit. **[2]**

Chapter

8 More waves and radiation

The following topics are covered in this chapter:

- ***Describing waves***
- ***Sound and ultrasound***
- ***Wave properties***
- ***Electromagnetic waves***

8.1 Describing waves

Motion of particles in waves

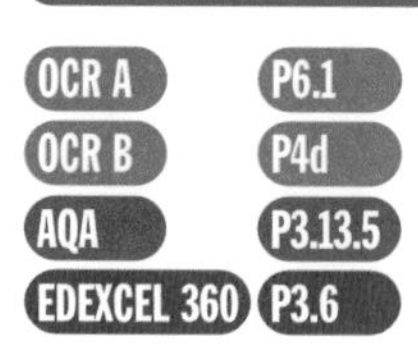

When a wave travels through a **medium**, the particles move backwards and forwards about their normal position – they **oscillate**. Fig. 8.1 shows this for a transverse and a longitudinal wave.

The **wavelength** is the distance between any particle and the next particle that is at the same stage in its oscillation, or cycle.

The **frequency** is the number of oscillations (or cycles) per second.

The **period** is the time for one complete oscillation (or cycle).

KEY POINT

For a wave, or other repeating motion

$$\text{frequency (Hz)} = \frac{1}{\text{period (s)}}$$

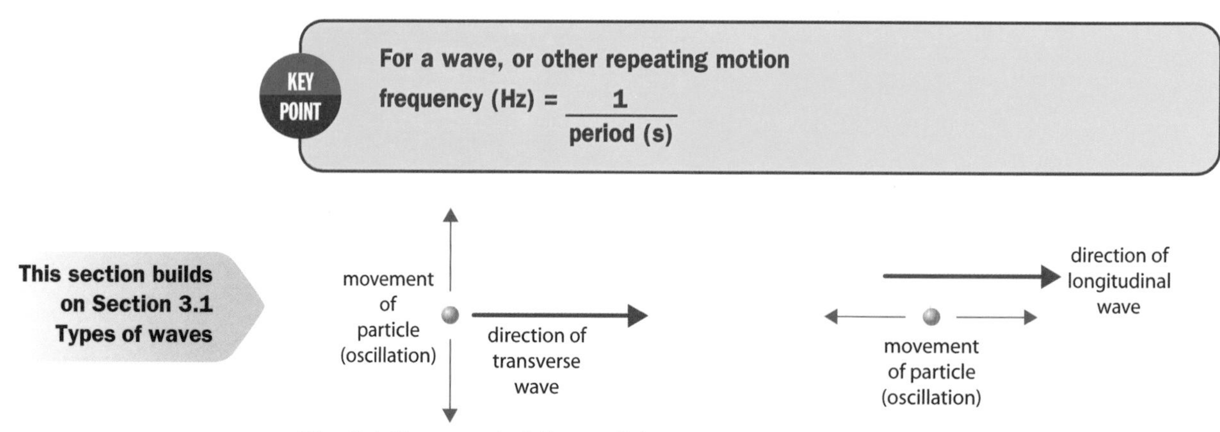

Fig. 8.1 Movement of the particles as a wave passes.

8.2 Sound and ultrasound

Sound waves

P3.13.5

This section builds on section 3.3 Sound and ultrasound.

Sound is caused by mechanical vibrations and travels as a wave through the material (solid, liquid or gas). **Sound** cannot travel through a **vacuum**.

The human ear

Sounds with **frequencies** in the range **20 Hz** to **20 000 Hz** can be detected by the human ear. The sound you hear has a:

- **pitch** which increases with the **frequency** of the sound wave
- **loudness** which increases with the **amplitude** of the sound wave
- **quality** which depends on the **waveform** – the shape – of the sound wave.

Looking at sound waves

An **oscilloscope** is used to look at the shape of sound waves. A microphone converts the sound wave to an electrical signal so that the voltage of the signal varies in time in the same way as the amplitude of the sound wave. The electrical signal is connected to the input of the **oscilloscope**, which displays a **graph** of the **amplitude against time**. Fig. 8.2 shows an oscilloscope being used to display the waveform for a sound with frequency 132 Hz.

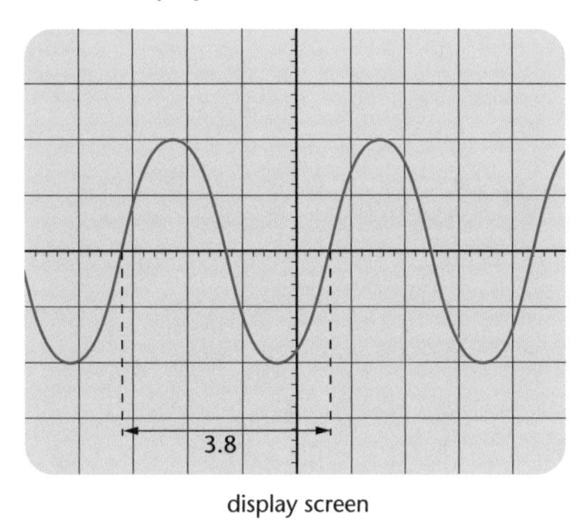

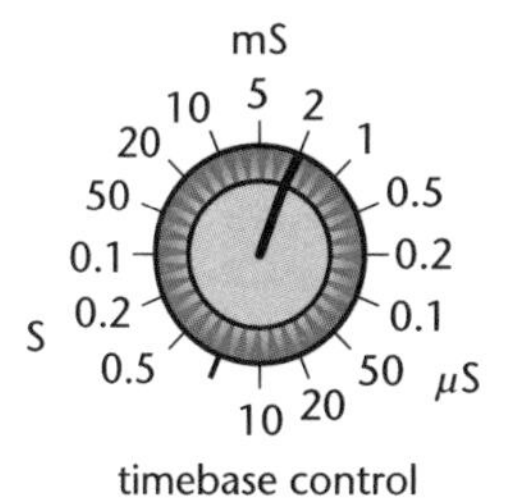

Fig. 8.2 An oscilloscope displaying a frequency of 132 Hz.

An oscilloscope can be used to find the period and frequency of an a.c. supply.

On the display

1 cycle = 3.8 divisions on the horizontal scale

The oscilloscope is set to T/div = 2 ms (one division = 2 ms = 0.002 s)

So the **period** (time for 1 cycle) = 3.8 × 2 ms = 7.6 ms

$$\text{frequency} = \frac{1}{\text{period}} = 132\,\text{Hz}$$

Sound waves and electromagnetic waves – the differences

OCR A P6.3

You will not be expected to remember the speed of electromagnetic waves in the exam.

Sound waves	Electromagnetic waves
longitudinal waves	transverse waves
cannot travel through empty space (a vacuum)	can travel through empty space (a vacuum)
can travel through solids, liquids and gases	can travel through some solids, liquids and gases
oscillations (movement back and forth) of the particles in the medium	oscillations (movement from side to side) of electric and magnetic fields
travel about a million times slower than electromagnetic waves in air	travel at 300 000 km/s in a vacuum, very slightly slower in air

Ultrasound

Ultrasound waves are sound waves with a **frequency** that is too high for humans to hear. The frequency is above the **upper threshold** of human hearing.

Ultrasound is also used to **scan** parts of the body, like the eye or an unborn fetus. This works because part of the pulse is reflected at each **boundary between** different layers of tissue. The reflections from the tissue boundaries are all used to build up a picture.

Ultrasound scans have two advantages over X-ray scans:

- ultrasound does not damage living cells or DNA and does not cause mutation
- ultrasound can produce images of soft tissue.

Ultrasound is also used to break down stones such as kidney stones or gall stones. The vibrations caused by the ultrasound make the stone break down.

Using ultrasound

AQA P3.13.6

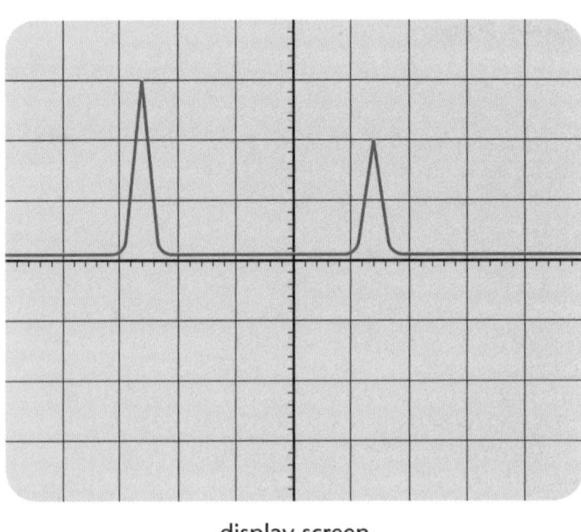

Fig 8.3 Measuring distance with ultrasound.

This diagram shows how the distance to the boundary between two different media can be detected using ultrasound. At the boundary an ultrasound wave is **partially reflected** – the rest is transmitted. The probe emits a pulse of ultrasound waves, and detects the returning pulse. The two pulses are displayed on an oscilloscope.

In this example, the distance between the pulses on the display is 4.0 divisions

The oscilloscope is set to 20 μs per division

So the time for the pulse to travel to the boundary and back is
$4.0 \times 20\,\mu s = 80\,\mu s$.

The time for the pulse to travel to the boundary = $\frac{1}{2} \times 80\,\mu s = 40\,\mu s$

The average speed of ultrasound waves in the body is 1500 m/s

Remember to change milliseconds to seconds.
For how to re-arrange formula see page 225.

distance = speed × time

$$\text{distance} = 1500\ \text{m/s} \times \frac{40\ \mu s}{1000\,000} = 0.06\ \text{m} = 6\ \text{cm}$$

This method can be used for **quality control**, for example to find cracks in railway tracks.

In **pre-natal scanning** a computer detects all the reflections and builds up a picture of the fetus.

8.3 Wave properties

More about light and sound

These 'More about' sections build on 3.5 Wave properties.

Reflection

Sound can be reflected, and the **angle of incidence** equals the **angle of reflection** as shown on page 47. Reflections of sound waves are called **echoes**.

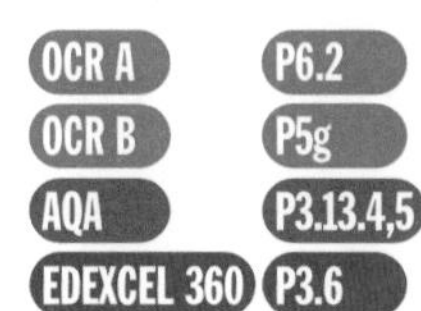

Refraction

When the **speed** of a wave changes, there is a change in the **wavelength**, because there is no change in the **frequency**. This can cause a change in direction as shown on page 48. Light travels more slowly in glass than in air, so a light ray is refracted as it passes through a glass block as shown in Fig. 8.4.

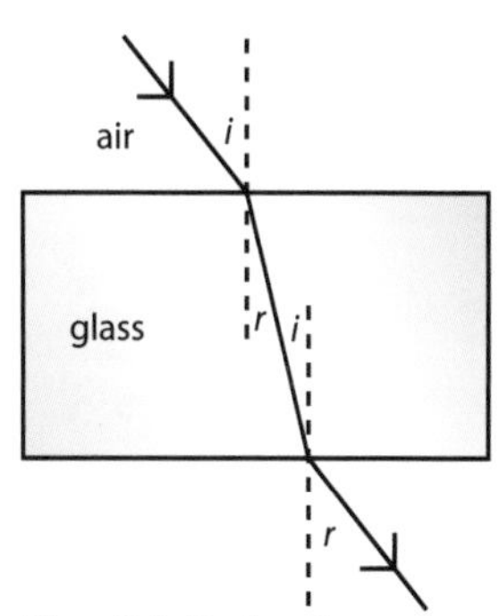

Fig. 8.4 Refraction of light in a glass block.

Sound is refracted, for example, in a different gas. Fig. 8.5 shows how a balloon full of **carbon dioxide** will **refract** sound waves (they travel more slowly in carbon dioxide than in air). More sound reaches the ear with the balloon in place than without the balloon.

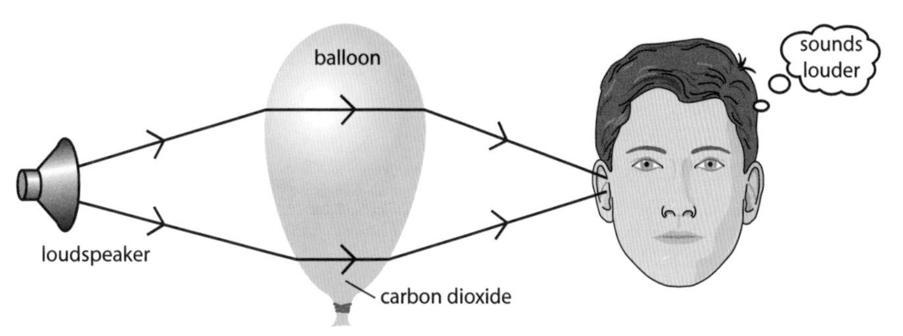

Fig. 8.5 Refraction of sound through a carbon dioxide filled balloon.

Dispersion

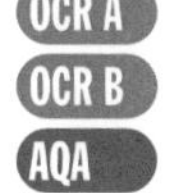

When light enters a different medium the change in speed and wavelength depends slightly on the frequency. The higher the frequency the greater the change in speed. This effect is called **dispersion**. For example, when white light passes from air to glass, the highest frequencies, (the violet end of the

Remember which colour bends the most by: Violet Veers Violently.

spectrum) are refracted towards the normal more than the lower frequencies (the red end of the spectrum.) When the light passes from glass to air the higher the frequencies the more the light is refracted away from the normal.

In a shape like a triangular **prism**, as shown in Fig. 8.6, this results in the white light being spread into a spectrum.

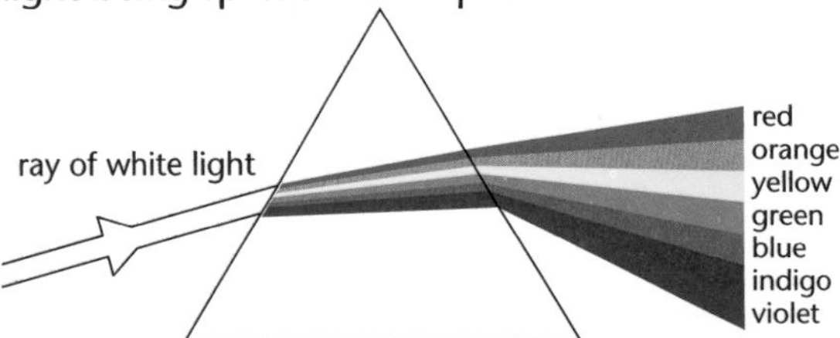

Fig. 8.6 Dispersion by a glass prism.

Refractive Index

OCR B P5e

KEY POINT

The refractive index, n, of a medium = $\dfrac{\text{speed of light in a vacuum}}{\text{speed of light in the medium}}$

refractive index is a ratio, it has no units

The angle of refraction when light enters a medium depends on the refractive index.

For how to rearrange formulae see page 225.

KEY POINT

Snell's law

$$n = \frac{\sin i}{\sin r}$$

where n is the refractive index, i is the angle of incidence and r is the angle of refraction

In real situations light passes from one **medium** to another, not from a vacuum to a medium. The **refractive index** for this transition is

The **refractive index**, $n = \dfrac{\text{speed of light in the incident medium}}{\text{speed of light in the refracting medium}}$

and n depends on the refractive indices of the two media.

KEY POINT

For an interface between two materials the refractive index, n, is

$$n = \frac{n_r}{n_i}$$

where n_r is the refractive index of the refracting medium and n_i is the refractive index of the incident medium

Fig 8.7 Light refracts when passing from the incident medium to the refracting medium.

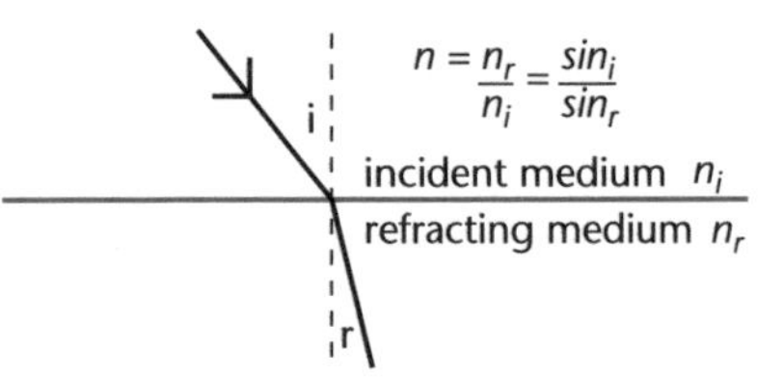

Dispersion

In the previous section, dispersion was explained by saying that the change in speed depends on the frequency of the light. Another way of saying this is to say that the **refractive index depends on the frequency of the light**.

Total Internal Reflection

The conditions for TIR are covered in section 3.5.

The **critical angle** is the **angle of incidence** that gives an **angle of refraction** of **90°**. Total internal reflection only happens when light passes from a dense to a less dense medium, so that the angle of refraction is larger than the angle of incidence. The angle of refraction depends on the **refractive indices** of the two materials. The larger the difference, the more the light refracts, and the lower the **critical angle**.

> **KEY POINT**
>
> **The critical angle, c can be calculated by using the equation**
>
> $$\sin c = \frac{n_r}{n_i}$$
>
> **where n_i is the refractive index of the incident medium**
> **and n_r of the refractive index of the refracting medium**

When comparing media in air, the higher the refractive index the lower the critical angle. Diamond has a very high refractive index and so a low critical angle, this leads to more internal reflections which is why diamonds sparkle so much.

More about light and sound – Diffraction

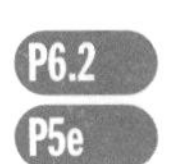

OCR A P6.2
OCR B P5e

> **KEY POINT**
>
> **For diffraction to occur at all, the gap must be of a similar size to the wavelength of the wave.**

The wavelength of sound is about a metre, so it is diffracted through doorways. (It is also reflected off walls, so hearing around a corner is partly due to diffraction, and partly reflection.) For light to be diffracted, the gap size must be about a thousandth of a millimetre – because the wavelength of light is so small. Patterns caused by diffraction of light can be seen through materials with a fine mesh, for example, a street lamp seen through a net curtain.

More about light and sound – Interference

OCR A P6.2
OCR B P5f

When identical sound waves from two speakers meet, there are positions where the waves arriving reinforce and give louder sound, and other positions where the waves cancel out to give a quiet point. Fig. 8.8 shows the pattern of loud and quiet areas.

A famous example of an **interference pattern** formed by light waves is called Young's experiment. Fig 8.9 shows how two very narrow slits close together are illuminated and the **diffracted light** from each slit overlaps and produces an interference pattern of **bright** and **dark lines**.

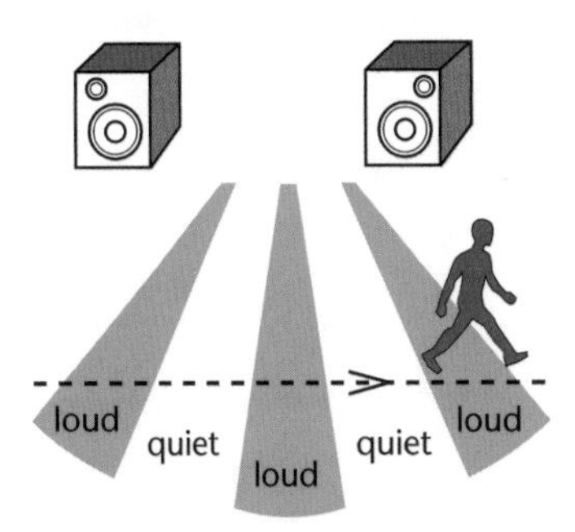

Fig. 8.8 Interference of sound waves.

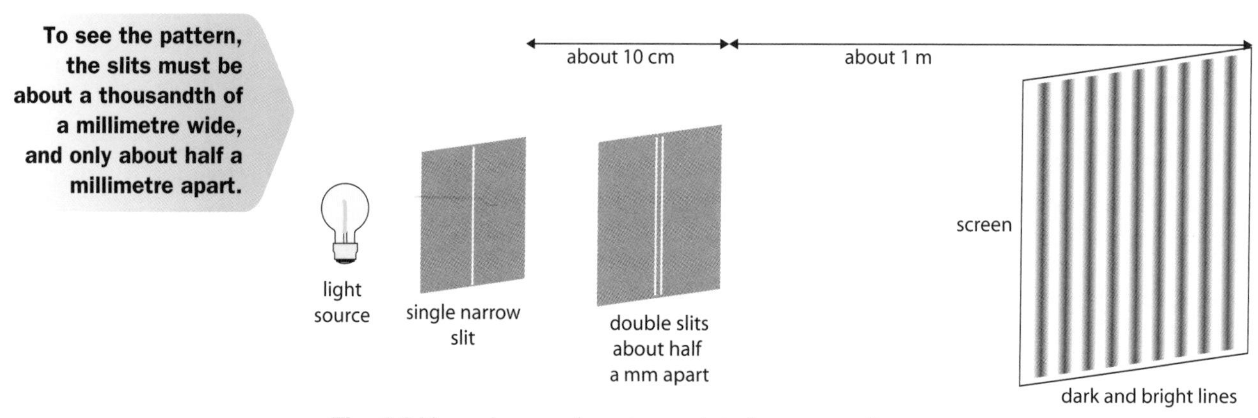

Fig. 8.9 Young's experiment – an interference pattern.

Path difference

OCR B P5f

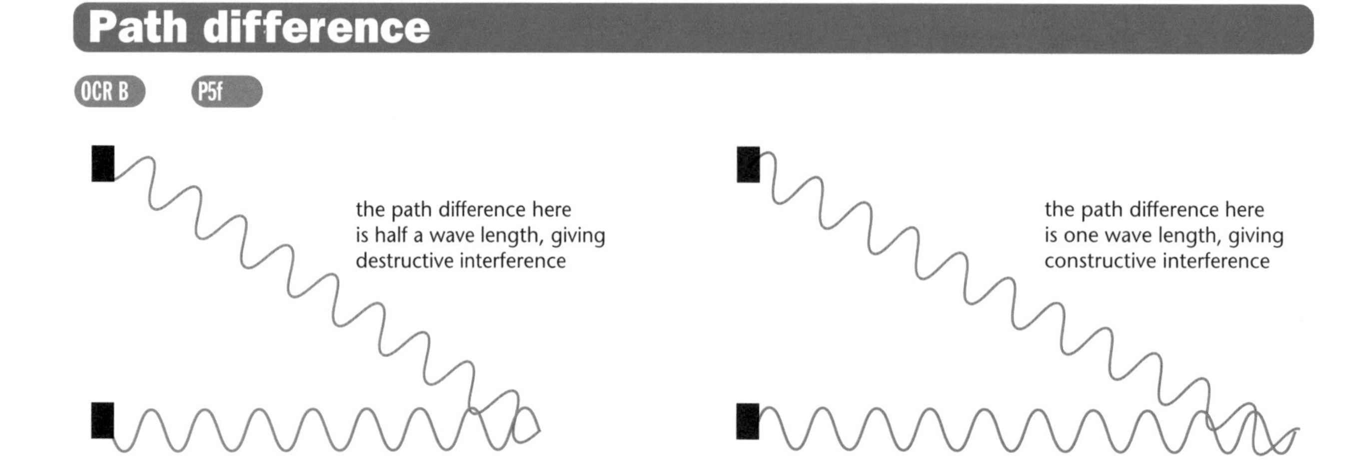

Fig 8.10 How different path difference gives an interference pattern.

When the **path difference** between the length of the paths taken by waves from two sources is equal to a **whole number** of **wavelengths** the waves arrive in phase (in-step) and result in **constructive interference**.

But when the **path difference** is equal to an **odd number** of **half wavelengths** the waves arrive out of phase (out-of-step) and result in **constructive interference**.

Evidence that light and sound are waves

Particles can be reflected. For example, on a snooker or pool table a ball striking the cushion bounces off so that the angle of incidence equals the angle of refraction.

Particles can be refracted. For example, if a ball rolls along a flat surface towards a ramp, but not at right angles to the ramp, when it reaches the ramp it will change direction. Fig. 8.11 shows the path of the ball viewed from above.

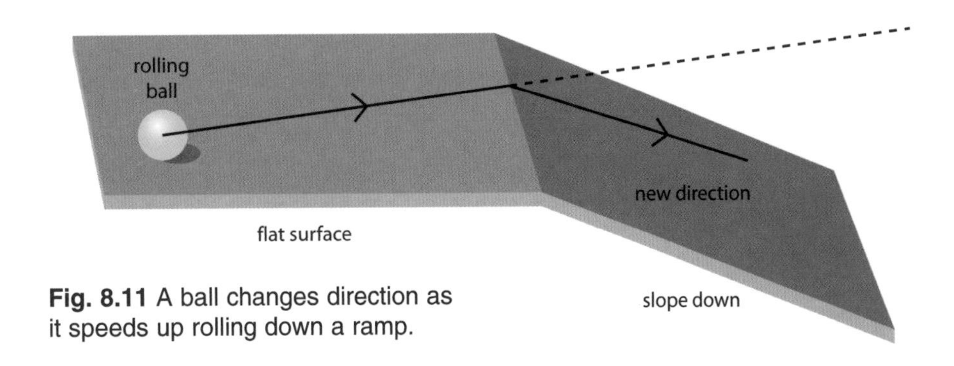

Fig. 8.11 A ball changes direction as it speeds up rolling down a ramp.

KEY POINT **Particles cannot be diffracted, or show interference effects. These properties are evidence of wave behaviour.**

Property	Light?	Evidence that light is a wave?	Sound?	Evidence that sound is a wave?
reflection	yes	no – could be a particle	yes	no – could be a particle
refraction	yes	no – could be a particle	yes	no – could be a particle
diffraction	yes	yes	yes	yes
interference	yes	yes	yes	yes

Polarisation

OCR B P5f

In transverse waves, the vibrations are at right angles to the direction of travel. This means that they can be up and down, or from side to side, or any angle in between. Fig. 8.12 shows a number of possible angles. A source of light will usually contain vibrations at all the possible angles. Transverse waves can be **polarised** so that all the **vibrations** are in the **same plane**. Fig. 8.12 shows a polarised wave with all the vibrations contained in a horizontal plane. This is called a **plane polarised** wave. One way to polarise light waves is by using a material called **Polaroid**. This is used to make **Polaroid sunglasses**. The sunglasses **reduce** light **intensity**, because some of the vibrations are blocked by the **Polaroid filter**. They also **reduce reflections** from horizontal surfaces, like water, because the reflected light is partly polarised in the horizontal plane and the Polaroid filters in the sunglasses are lined up to block light in this plane.

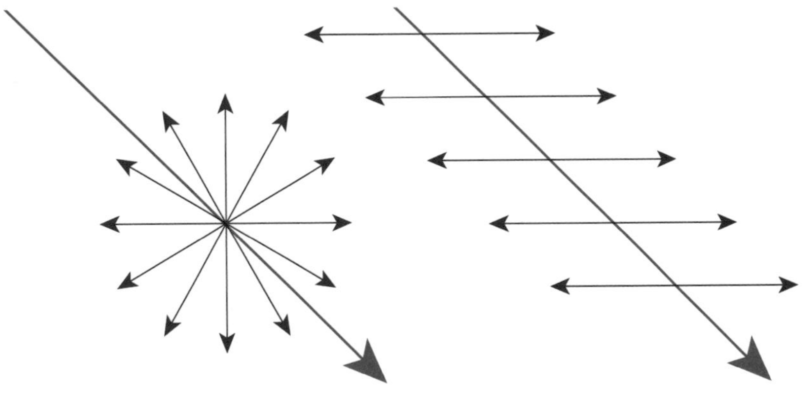

Fig 8.12 An unpolarised wave and a horizontally plane polarised wave

8.4 Electromagnetic waves

The photon model

Electromagnetic radiation can be described as packets of **energy** called **photons**.

The energy delivered by each photon in a beam of electromagnetic radiation increases with the frequency of the radiation.

For example, a gamma ray photon has a higher frequency, and photon energy than a microwave photon.

The **intensity** of a beam of electromagnetic radiation is the **energy** it delivers each second. The intensity depends on the **number of photons** arriving each second, and the **energy** carried by **each photon**.

X-rays and gamma rays

EDEXCEL 360 P2.11

The difference between gamma rays and X-rays is in the way they are produced. **Gamma rays** come from the nucleus of some radioactive atoms. **X-rays** are produced by firing high speed electrons at a metal target in an X-ray machine. X-rays are easier to control than gamma rays. **X-rays** are used for:

- **Medical scans**, for example to show broken bones. They pass through the body tissues but are stopped by denser materials such as the bones, which show up as shadows on film or a screen.
- **Security scans** of passengers' luggage. They pass through the suitcase and clothes, but metal items and batteries stop the X-rays and show up as shadows on the screen.

Uses of gamma rays are covered in section 1.1 Radioactivity

More about radio waves and microwaves

Remember (sections 3.1 and 3.4), all electromagnetic waves travel at 300 000 000 m/s in a vacuum, or in space, and only slightly slower in air.

Both radio waves and microwaves are used for communications, but they behave differently depending on the wavelength and frequency of the waves.

There are two major effects that affect the choice of wavelength:

- **diffraction** through gaps, around obstacles, and from the transmitter
- **absorption** and **scattering** by the atmosphere, including dust and water vapour in the atmosphere, and also by water droplets in rain and clouds.

Wavelength and frequency are related by the wave equation, $v = f\lambda$.

Microwaves (wavelengths of a few centimetres) are used to transmit signals to and from orbiting **satellites**. Diffraction causes the signal to spread out from the transmitting satellite dish, but the dish can easily be made bigger than the wavelength so the diffraction effects can be reduced, (remember the maximum diffraction occurs when the gap size – in this case the dish size – equals the wavelength). So, compared to radio waves, microwaves can be sent as a **thin beam**.

Radio waves have longer wavelengths and are **diffracted** between **buildings** and **around hills**, so they reach all areas and are suitable for broadcasting. **Long wave radio** (wavelengths of about a kilometre) is also diffracted by the curved shape of the Earth, over the **horizon**. This means they have a very long range.

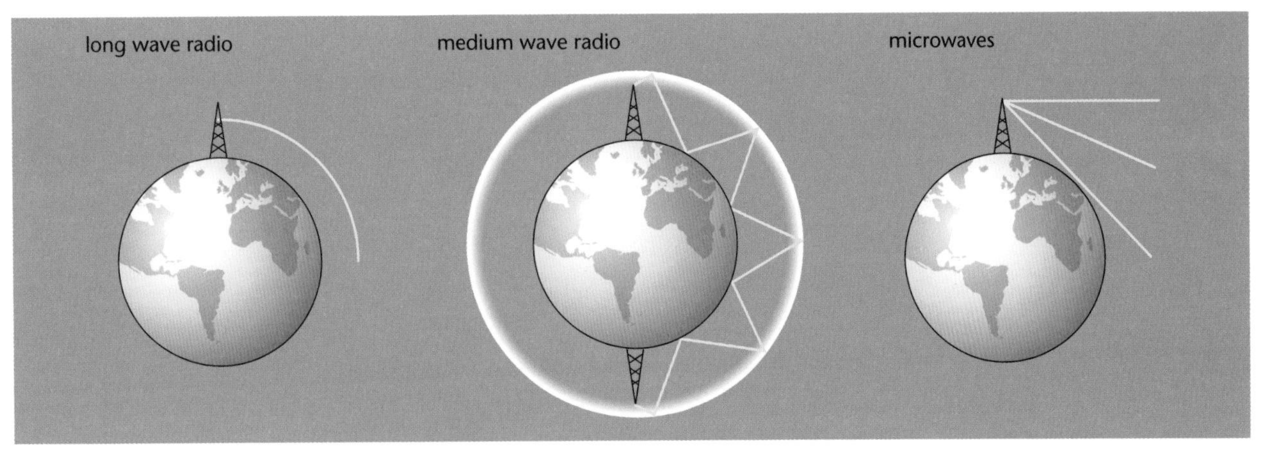

Fig 8.13 How radio waves and microwaves travel.

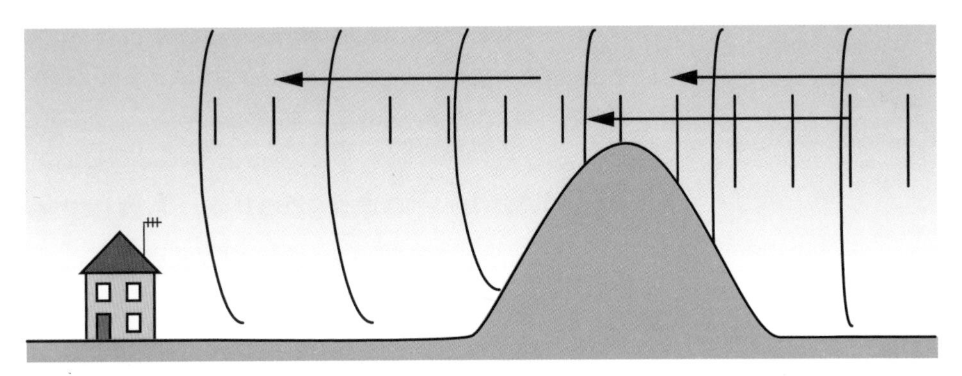

Fig 8.14 Diffraction is the reason that this house receives long wavelength broadcasts but not short wavelength broadcasts.

For OCR B you are expected to remember the information in red in the table.

Table 8.1 summarises the behaviour of microwaves and radio waves at different wavelengths and frequencies.

Range		Typical wavelength	Frequency range	
Radio waves (below 1 GHz)	**Long wave**	1 km	30 kHz – 2 MHz	Long range, diffracted by hills and horizon.
	Medium wave	100 m – 10 m	3 MHz – 30 MHz	Diffracted by hills. Below 30 MHz reflected by the ionosphere
	Short wave and above	1 m – 10 m	30 MHz – 300 MHz	
Microwaves (Between infrared and radiowaves)		1 cm	3 GHz – 300 GHz	Above 30 GHz rain and dust reduce signal strength by absorption and scattering.
Infrared (above 300 GHz)		1 mm	300 GHz	

Modulation

 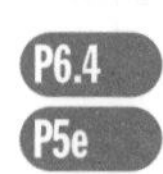

OCR A P6.4
OCR B P5e

For communications with light see section 4.4 Infrared and light.

To communicate information, for example music or data, the information must be added to the light or radio wave. For example, these are the steps to broadcast music using analogue radio:

- A microphone is used to produce an electrical signal that matches the music (an analogue signal). This is called an **audio frequency (AF) signal**
- A radio wave frequency band (small range of radio frequencies) is chosen to carry the signal. This is called a **radio frequency (RF) carrier**
- The **AF signal** is added to the **RF carrier** in a process called **modulation** (The AF signal is used to modulate the RF carrier)
- The modulated radio wave is broadcast from a transmitter aerial
- The radio receiver aerial receives all the transmitted radio waves. Its job is to reproduce the original sound – the music. You tune the radio receiver to the **RF carrier** you want – the one used by the radio station broadcasting the music
- The modulated radio wave is received and demodulated to remove the **RF carrier** from the **AF signal**
- The **AF signal** will have decreased in intensity and needs amplifying. It is amplified and sent to the loudspeaker.

Amplitude modulation and frequency modulation

KEY POINT

There are two types of modulation. In amplitude modulation (AM) the amplitude of the radio wave is varied to match the audio frequency (AF) signal. In frequency modulation (FM) the frequency of the radio wave is varied to match the audio frequency (AF) signal.

For OCR B you only need to know about AM.

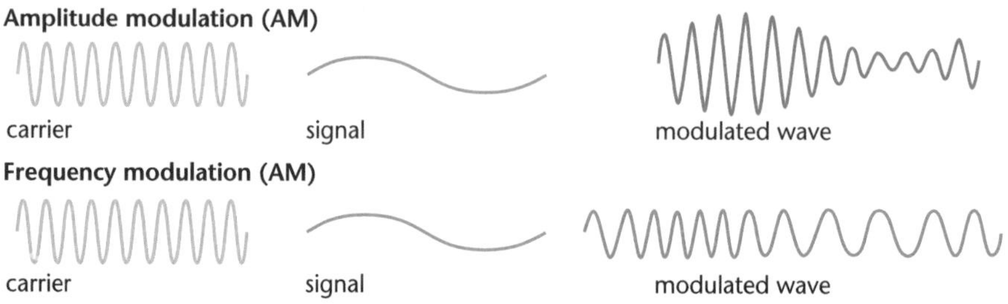

Amplitude modulation (AM)

carrier signal modulated wave

Frequency modulation (AM)

carrier signal modulated wave

Fig. 8.15 Amplitude modulation (AM) and frequency modulation (FM)

As the signal travels it picks up random additional signals – called **noise**. This reduces the quality of the signal. When the signal is amplified the noise is amplified too, as shown in Fig. 8.16. The amplitude is more affected by noise than the frequency, so **FM** broadcasts are **less affected** by **noise** than **AM** broadcasts.

Digital signals

Digital radio has an extra step involved. The AF signal is converted to a **digital AF signal** which is then used to modulate the radio frequency (RF) carrier wave. The digital AF signal must be decoded in the radio receiver before it is sent to the loudspeaker. The advantage, shown in Fig. 8.16, is that the 'on' and 'off' states can still be recognised despite any **noise** that is picked up. The signal can be 'cleaned up' to remove any noise so digital signals can be **regenerated** and give higher quality reception.

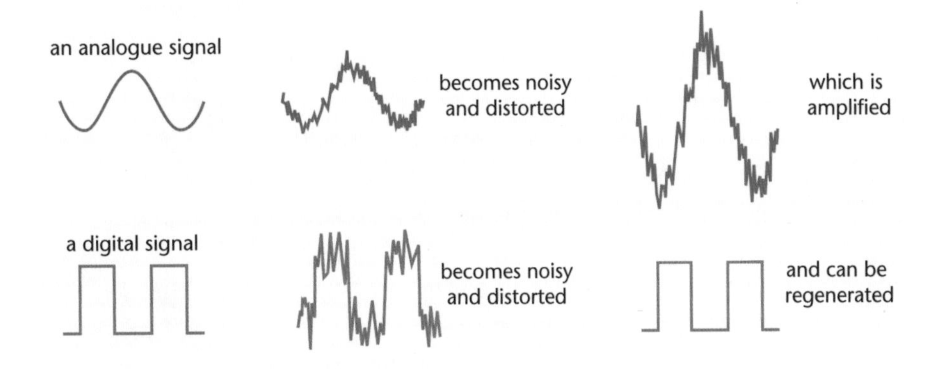

Fig. 8.16 Digital transmission can have higher quality.

HOW SCIENCE WORKS

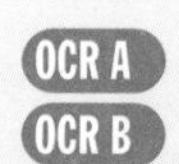

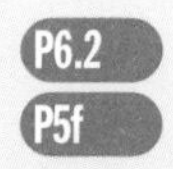

What is light?

The boxes A to H describe the stages in the development of scientists' theories about light. Box A is the first box. Read them all and see if you can decide the correct order for the others.

Box A

If it is completely dark and you have a searchlight you can move it around and pick out objects that you can see in the light beam. Most children begin with this idea of how the eye works – it sends out a ray to see things. It is possible this was the earliest idea of how light behaved.

Fig. 11.22 Seeing an object?

Box B

In 1704, Isaac Newton did some experiments with light. He knew that waves were diffracted when they passed through small gaps. He checked very carefully and could see no evidence that light was diffracted. He said light must be a stream of particles.

Fig. 11.23 Shadows have sharp edges.

Box C

In 1905, Albert Einstein published his theory that light behaves as a beam of particles called photons. He used this theory to explain why high frequency light caused electrons to be emitted from the surface of some metals when low frequency light did not – no matter how much you turned up the intensity of the light.

Box D

About 2500 years ago, the Ancient Greeks suggested a theory that light was a stream of particles, for example from the Sun or a lamp. When the particles bounced off objects the light was reflected. When people saw objects it was because the light particles had entered their eyes.

HOW SCIENCE WORKS

Box E

Today scientists have accepted that sometimes light behaves as a wave and sometimes it behaves as a particle. They use a wave theory of light or a particle theory of light depending on which is best suited to the situation they are trying to explain.

Box F

In the fifteenth century Leonardo da Vinci compared reflections of light to echoes of sound. As sound was considered to be a wave, this suggested that light was also a wave.

Box G

Thomas Young's ideas were accepted in other parts of Europe before they were accepted in the United Kingdom because British scientists had such a high opinion of Isaac Newton, they did not believe his ideas could be wrong.

Box H

In 1802, Thomas Young showed that – with small enough gaps – light does show diffraction and interference effects that can only be explained by a wave theory. The gaps had to be about the width of a hair because the wavelength of light is so small.

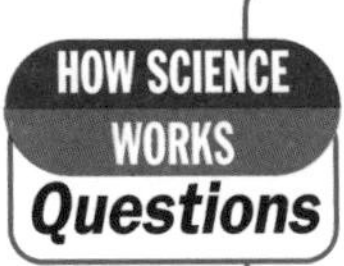

1. What is the correct order of the boxes? **[4]**
2. There are several reasons why scientists sometimes find it difficult to accept new explanations. Why did some UK scientists find it difficult to accept that light was a wave? **[1]**
3. Today scientists say that light is a beam of particles called photons. Does this mean that Thomas Young was wrong to say that light was a wave? Explain your answer. **[2]**

Exam practice questions

1. Light strikes a glass surface at an angle of 40° and is refracted. The refractive index of the glass is 1.5. What is the angle of refraction?

Use $n = \frac{\sin i}{\sin r}$

A 25°
B 40°
C 43°
D 90° **[1]**

2.

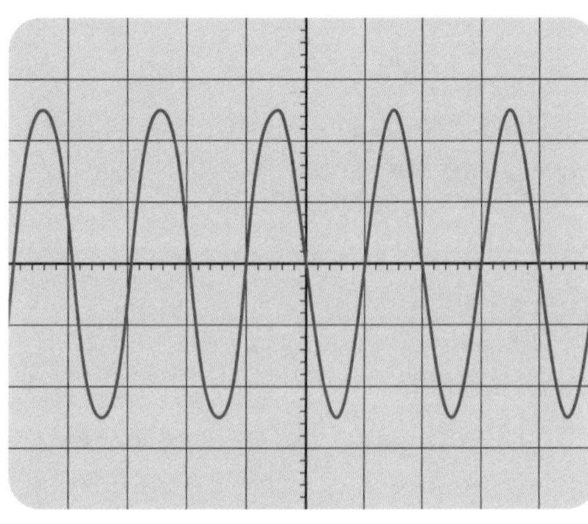

display screen

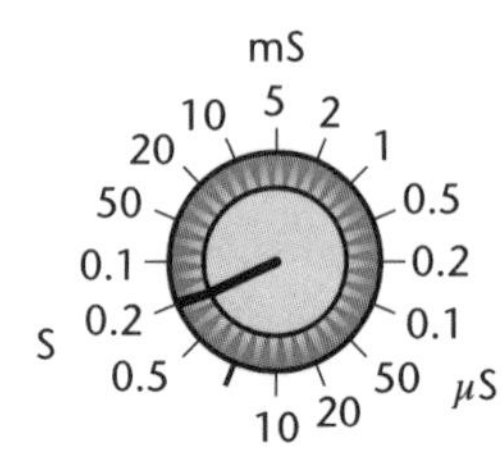

timebase control

The period of the wave displayed on the oscilloscope is

A 0.2 seconds
B 0.4 seconds
C 2.0 seconds
D 2.5 seconds **[1]**

3. A digital signal can be transmitted with higher quality than an analogue signal because:

A The radio wave does not have to be modulated
B The digital signal does not pick up so much noise
C The digital signal does not need to be amplified
D The digital signal can be cleaned up and then amplified **[1]**

4. Some ocean waves have a wavelength of 140 m and 1 wave passes a rock every 20 seconds. The distance from the top of a crest to the bottom of a trough is 3 m.

Tick ✔ **true** or **false** for each statement.

	True	**False**
The period of the waves is 20 seconds		
The frequency of the waves is 20 Hz		
The wave speed is 7 m/s		
The amplitude is 3 m		

[4]

Exam practice questions

5. Choose words from the list to complete this description of how radio waves can be used to broadcast a TV programme. (Use words once, more than once, or not at all.)

aerial **analogue** **carrier** **demodulated**
digital **modulate** **signal** **transmitter**

The TV signal is used to ____1____ the radio frequency ____2____ wave. The radio wave is broadcast from the ____3____ aerial. The receiver ____4____ picks up the broadcast ____5____. The radio wave is ____6____ to recover the TV signal. **[6]**

6. Draw lines to join the range of electromagnetic waves to the use.

Range	**Use**
microwaves	communication with satellites
radio waves	optical fibre communications
infrared	medical scans
x-rays	TV broadcasts

[3]

7. **(a)** What is ultrasound?
(b) Ultrasound is used to build up an image of the unborn fetus. Explain what happens to the ultrasound when it reaches the fetus.
(c) State another medical use, other than scanning parts of the body, of ultrasound. **[4]**

8. **(a)** Complete the labels to describe the sound wave shown in the diagram below.

A ______________

B ______________

C ______________

D ______________ wave

(b) A transverse wave moves from left to right across the water. Describe how the particles move as the wave passes. **[6]**

Chapter

9 Nuclear physics

The following topics are covered in this chapter:

- ***Atomic structure***
- ***Changes in the nucleus***
- ***Nuclear fusion***

9.1 Atomic structure

A model of the atom

Before 1910 scientists had a **plum pudding** model of the **atom**. They imagined that the atom was made of positively charged material (the pudding) with the negatively charged electrons distributed inside it (the plums). Ernest Rutherford realised that **alpha particles** were smaller than atoms and thought up an experiment to find out more about the structure of atoms. Alpha particles were fired at a very thin sheet of **gold foil**, rolled out to be just a few atoms thick. The experiment was done in a vacuum, so that the alpha particles were not stopped by the air. It was surrounded by a screen made of fluorescent material, so that a small flash of light was seen when an alpha particle hit the screen. Hans Geiger and Ernest Rutherford did the experiment – they counted small flashes of light at different angles for hours. The table shows their observations.

What happened to the alpha particles	Ernest Rutherford's explanation
Most went straight through the foil, without being deflected	The atom is mostly empty space
Some were deflected, and there was a range deflection angles	Some passed closer to charged parts of the atom
A small number were 'back-scattered' – they came straight back towards the alpha particle source	There are tiny regions of concentrated positive charge which repel the small number of alpha particles that have a head-on collision

Rutherford said the back-scattering surprised him – it was as if you fired a gun at a piece of tissue paper and the bullet came back at you.

Fig. 9.1 shows how this model accounts for the observations.

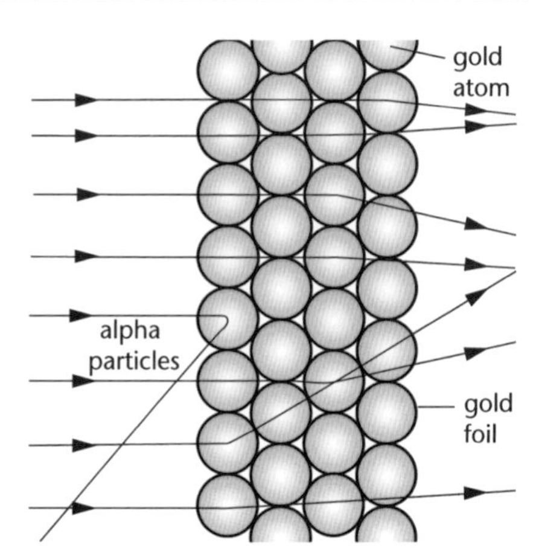

This lead to our 'solar system' or **nuclear model** of the atom.

- The atom has a central nucleus where most of the mass is concentrated. It is positively charged.
- The nucleus is very small compared to the volume, or shell, around the nucleus that contains the electrons.

Different atoms

AQA P2.12.9
EDEXCEL 360 P2.11,P3.5,P3.6

This table summarises the properties of the particles found in the atom. The values for relative mass and the relative charge give a comparison of the particles.

The protons and neutrons are collectively called nucleons.

Particle	Symbol	Where found in the atom	Relative mass	Relative charge
proton	p	in nucleus	1	+1
neutron	n	in nucleus	1	0 (neutral)
electron	e	outside nucleus	$\frac{1}{1840}$	–1

KEY POINT
The atomic number or proton number, Z, is the number of protons in the nucleus. Atoms of different elements have different numbers of protons.

The mass number or nucleon number, A, is the total number of protons and neutrons in the nucleus. Atoms of the same element with different numbers of neutrons are called isotopes.

The nucleus is given a symbol $^{A}_{Z}El$ where A is mass number Z is the proton number and El is the chemical symbol for the element. For example $^{12}_{6}C$ is a stable isotope of carbon with 6 protons and 6 neutrons.

9.2 Changes in the nucleus

Radioactive decay

OCR B P 4f
AQA P2.12.9
EDEXCEL 360 P2.11

A radioactive nucleus is unstable and emits nuclear radiation. This process is called **radioactive decay**. It is not possible to predict when this will happen, nor is it possible to make it happen by a chemical or physical process, (for example by heating it). The **decay** is **random**.

Radioactive emissions

OCR B P 4f
AQA P2.12.9
EDEXCEL 360 P3.6

Alpha emission is when **two protons and two neutrons** leave the nucleus as one particle. This alpha particle is identical to a helium nucleus so has the symbol: $^{4}_{2}\mathbf{He}$. The new nucleus has a mass number which has decreased by four and an atomic number which has decreased by 2. For example radon –220 is a gas that decays by alpha emission.

$^{220}_{86}Rn \rightarrow\ ^{216}_{84}Po +\ ^{4}_{2}He$

Beta emission occurs when a **neutron** decays to a **proton** and an **electron** and the electron leaves the nucleus. The atomic number increases by one and the mass number is unchanged. The beta particle is a high energy electron and has the symbol: $^{0}_{-1}e$

$^{14}_{6}C$ is a radioactive isotope of carbon with 6 protons and 8 neutrons.

This nucleus emits beta radiation and forms a nitrogen nucleus:

$^{14}_{6}C \rightarrow\ ^{14}_{7}N +\ ^{0}_{-1}e$

A beta particle is an electron from the nucleus – not an orbital electron.

Gamma emission occurs when the nucleus emits a short burst of high-energy electromagnetic radiation. The gamma ray has a high frequency and a short wavelength.

KEY POINT

alpha particle: a helium nucleus $^{4}_{2}He$
beta particle: high energy electron $^{0}_{-1}e$
gamma ray: electromagnetic wave.

9.3 Nuclear fusion

Nuclear fusion

The protons and neutrons inside the nucleus are held together by a force called the **strong force**. When two small nuclei are close enough together they can **fuse** together and form a larger nuclei. When they do this they release a large amount of energy. The problem is getting the two nuclei close enough for the strong force to take effect, because **nuclei** are positively charged and **repel** each other.

Inside **stars** the temperatures are high enough for the nuclei to have enough energy to get close enough for the strong force to take over and **nuclear fusion** occurs.

Scientific research is continuing to try and control the nuclear fusion reaction and produce nuclear fusion reactors.

HOW SCIENCE WORKS

OCR A P3.4
OCR B P4e,g
AQA P1.11.6
EDEXCEL 360 P2.9, P2.11

Radioactivity benefits and risks

In the USA in the 1920s, you could buy 'The Curie RadioActive Re-Generator and Stone Water Filter' Fig. 12.8. It was made by the Curie Radium Company and was sold as 'A water jar in which is placed your local drinking water. In this jar is also placed a Radium Ore Disc - this disc throws off light rays thereby forming niton gas, making the water radioactive, the same as the Great Health Springs. Radioactive water is a proven means to health as millions can testify.'

Today we would say the radioactive disc is made of uranium ore and gives off radon gas.

Radium glows in the dark so, from about 1917, radium-containing paint was used to paint luminescent hands and numbers on watches. Many workers were young girls. They were encouraged to use their lips to make the tip of the brush into a point. Radium-226 was absorbed into their bones. One girl reported that after blowing her nose the handkerchief glowed in the dark. After 1923 the link between radium dial painting and health problems began to be noticed, and the practice of using lips to point the brush, was discontinued around 1926.

In the 1970s radiotherapy for breast cancer was not as advanced as it is today. The dose had to be high enough to kill the cancer cells, or the cancer would return. In 2006 a woman died as a result of the large dose of radiation she received during a 20 week course of radiotherapy in 1972. The high dose was standard practice at the time. This woman lived for another 34 years but women who did not have the treatment died within a few years, when the cancer returned.

HOW SCIENCE WORKS

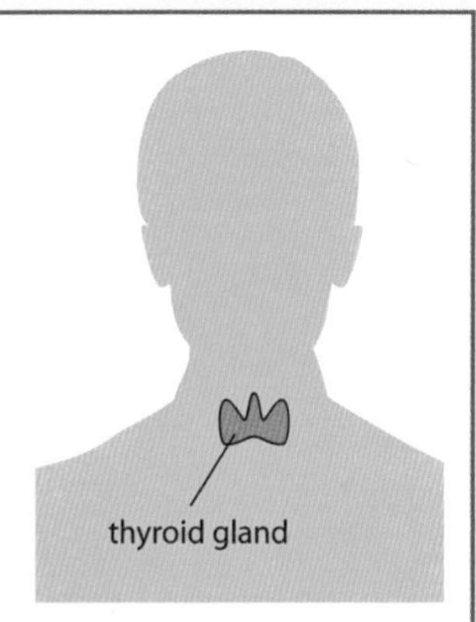

Jack had cancer of the thyroid gland. This gland in the neck absorbs iodine. To treat the cancer cells doctors gave Jack a drink containing radioactive iodine-131. For a few days beforehand he had to avoid eating and drinking products containing iodine, so that the radioactive iodine would be absorbed by his body. After taking the treatment Jack stayed in a side-room in the hospital. The visitors and nurses did not stay in the room very long because he was radioactive. Children were not allowed to visit him, and the nurse checked the radioactivity level with a Geiger counter. His urine was radioactive, so he had to flush the toilet several times to dilute it. Before he left, the books he had been reading were checked with a Geiger counter. After a few days the level of radioactivity had dropped to the normal background level and Jack was allowed to go home. The next check-up showed that the cancer cells had been killed by the radioactive iodine.

This table compares the risks of different activities by working out an average of days of life lost for each activity. This is worked out by:

$$\frac{\text{total days of life lost by all the people who died early}}{\text{total population}}$$

Activity	Average days/years lost
Smoking 20 cigarettes a day	6 years
All accidents	207 days
Cancer due to being exposed to 3 mSv of radiation (average normal background)	15 days
Cancer due to being exposed to 10 mSv of radiation	51 days

1. Explain whether you think the benefit is worth the risk.
   - **(a)** of using the The Curie RadioActive Re-Generator and Stone Water Filter
   - **(b)** of using the lips to get a point on a brush of radium-containing paint
   - **(c)** for the woman who had the 1970s breast cancer treatment
   - **(d)** of treating thyroid cancer with radioactive iodine **[4]**
2. Why is the The Curie RadioActive Re-Generator and Stone Water Filter not on sale today? **[2]**
3. One of Jack's friends, who is a smoker, avoided him after Jack left hospital because he was afraid Jack might be radioactive. Jack said that his friend was more at risk from smoking.
   - **(a)** Use the table above to explain why Jack was correct.
   - **(b)** Suggest why Jack's friend was more worried about radioactivity than smoking. **[2]**

Exam practice questions

1. $^{235}_{92}U$ and $^{238}_{92}U$ are both isotopes of uranium. Which of these tables is correct?

A

isotope	protons	neutrons
$^{235}_{92}U$	92	235
$^{238}_{92}U$	92	238

B

isotope	protons	neutrons
$^{235}_{92}U$	92	143
$^{238}_{92}U$	92	146

C

isotope	protons	neutrons
$^{235}_{92}U$	235	92
$^{238}_{92}U$	238	92

D

isotope	protons	neutrons
$^{235}_{92}U$	143	92
$^{238}_{92}U$	146	92

[1]

2. The half-life of technetium-99m is 6 hours. At the start of treatment the activity of a source of technetium-99m is 8000 counts per minute. After 24 hours the activity is:
   A 333 counts per minute
   B 500 counts per minute
   C 1333 counts per minute
   D 4000 counts per minute **[1]**

3. Factory waste is discharged from a pipe at sea. To check that it is not being washed up on a nearby beach, a radioactive tracer is added. A suitable isotope to use would be:
   A an alpha emitter with a half-life of 1 year
   B a beta emitter with a half-life of 2 days
   C a beta emitter with a half-life of 1 year
   D a gamma emitter with a half-life of 2 days **[1]**

4. Put these statements in order to describe how the age of a wooden spear is determined using carbon dating. Start with **C**.
   A A sample of the wood is tested to find the proportion of carbon-14
   B The proportion of carbon-14 in the wood falls as the nuclei decay
   C Carbon dioxide, containing some carbon-14, is taken in by the living tree during photosynthesis
   D The tree is cut down and no more carbon –14 is taken in
   E The result is compared with living wood and only half of the carbon-14 is left
   F The wood is made into a spear
   G The age of the spear is one half-life, which is 5730 years **[3]**

5. Write **T** for the **true** and **F** for the **false** statements below.
   (a) Nuclear reactors use fuel rods made of uranium –235 or plutonium-239.
   (b) The nucleus splits into two parts, this is called nuclear fusion.
   (c) A chain reaction occurs when a nucleus splits and releases a few neutrons which can be absorbed by other nuclei and cause them to split.

Exam practice questions

(d) Control rods are lowered into the nuclear reactor core to speed up the reaction.
(e) The energy released when the nuclei split heats up the reactor core. Coolant is circulated to remove the heat.
(f) The removed fuel rods contain radioactive isotopes with very long half-lifes. **[3]**

6. Use words from the list to complete the sentences below. (Use words once, more than once or not at all.)

alpha	**aluminium**	**beta**	**empty**	**gold**	**large**	**most**
negative	**none**	**positive**	**some**	**small**	**space**	

Hans Geiger and Ernest Marsden fired ____1____ particles at very thin ____2____ foil. They counted the small flashes of light as the particles hit fluorescent screens and discovered that ____3____ of the particles passed straight through the foil without being deflected. This showed that the atom was mostly ____4____ ____5____. A few of the particles were sent back towards the source. This showed that the atom had a ____6____ nucleus with a ____7____ charge. **[7]**

7. **(a)** Explain what is meant by **background radiation**.
(b) Write down one source of background radiation. **[2]**

8. Sodium–24 is a radioactive isotope of sodium that emits gamma radiation. The stable form of sodium is sodium–23.
a) Explain how the nuclear structure of sodium-24 is different to sodium-23 **[1]**

The activity of a sodium-24 source was measured over 2 days and this graph was plotted.

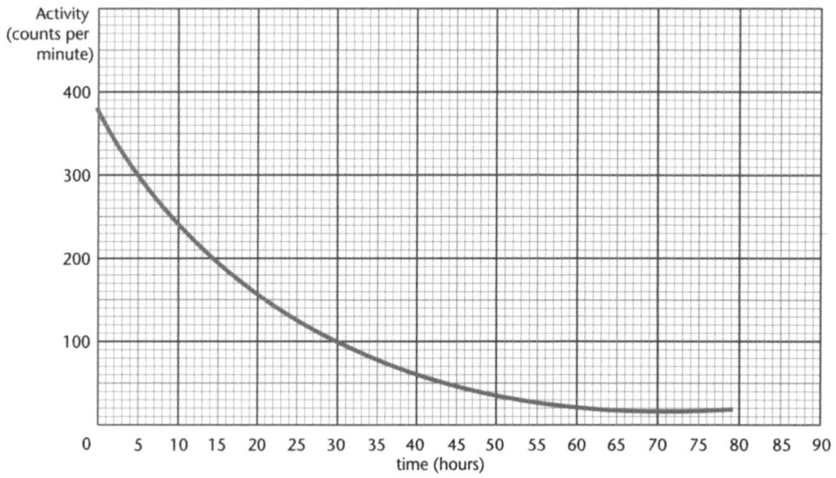

Activity (counts per minute)
400
300
200
100
0 5 10 15 20 25 30 35 40 45 50 55 60 65 70 75 80 85 90
time (hours)

(b) Use the graph to work out the half-life of sodium-24. **[2]**
(c) Use the graph to estimate the background activity at the site of the experiment. **[1]**
(d) Explain whether a salt of sodium-24 would be a suitable isotope to use for :
i) the source in a smoke detector
ii) a tracer to find the leak in an underground water pipe. **[4]**

Chapter 10

Particles

The following topics are covered in this chapter:

- *Properties of gases*
- *Atoms and nuclei*
- *Electrons and electron beams*
- *Fundamental and other particles*

10.1 Properties of gases

Gases

OCR A P 7.4
EDEXCEL 360 P3.5

Mass and volume

The gas in a closed container, for example, a balloon, has **mass**, measured in kilograms (kg) and **volume**, measured in metres cubed (m^3). The mass is fixed, but by squeezing the balloon you can change the volume.

Pressure and volume

A gas is made up of a very large number of atoms or molecules that are constantly moving. The movement of each of these particles is **random** – it changes speed and direction as it **collides** with other particles and the walls of the container. The forces exerted on the container surface by all the particles give the gas **pressure**.

A common mistake is to say that gas pressure results from collisions between particles.

KEY POINT

Gas pressure is equal in all directions and results from the forces exerted when the gas particles collide with the walls of the container.

When a gas is compressed into a **smaller volume** there will be more collisions with the walls, so the **pressure increases**, as shown in Fig. 10.1. Experiments show that halving the volume doubles the pressure.

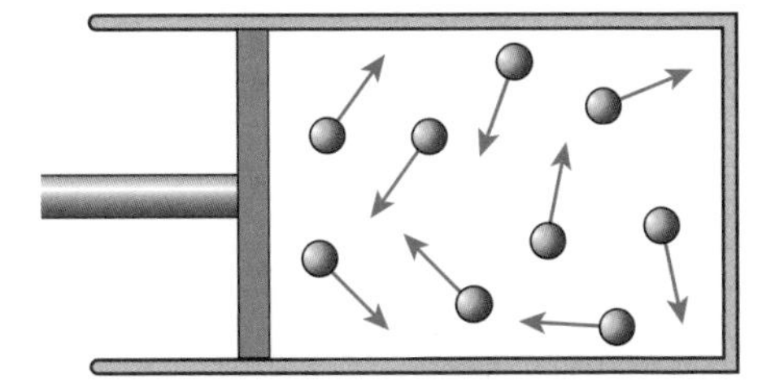

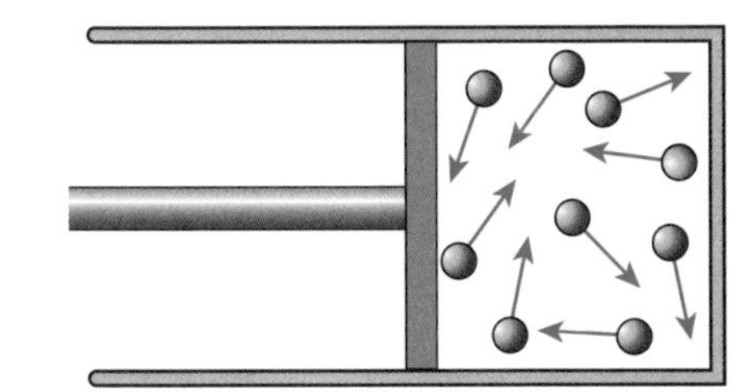

Fig. 10.1 Reducing the volume of a gas increases the pressure.

Pressure and temperature

Imagine a sealed glass container filled with gas, so the gas has a fixed mass and volume. When the gas is heated its **temperature** rises. The particles have more energy so their **average speed** is faster.

- A faster particle exerts more force during a collision.
- A faster particle will make more collisions.

These changes increase the pressure. When the **pressure** of a fixed volume of gas is plotted against its **temperature** the graph is a straight line, as shown in Fig. 10.2.

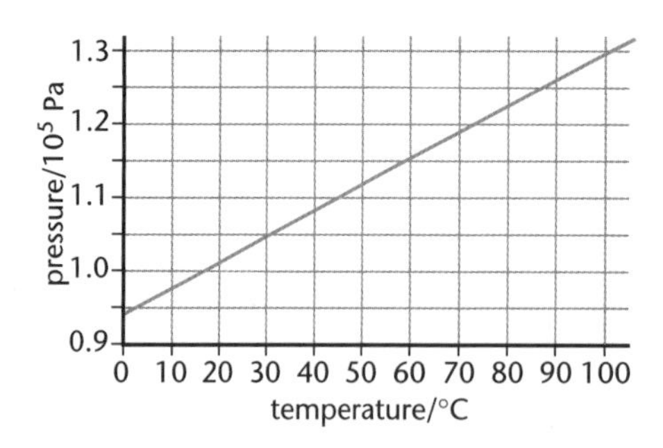

Fig. 10.2 Pressure of a gas increases with temperature.

Absolute zero temperature

A real gas would condense before it reached this temperature. This model imagines an ideal gas.

Notice that the line does not go through the point (0,0) because gases don't have zero pressure at 0°C. Extending the graph back to negative temperatures shows the temperature at which the gas would have zero pressure. This is the temperature where all its particles stop moving and make no collisions. It is called the **absolute zero** of temperature. It is not possible to get a lower temperature because it is not possible to get any colder.

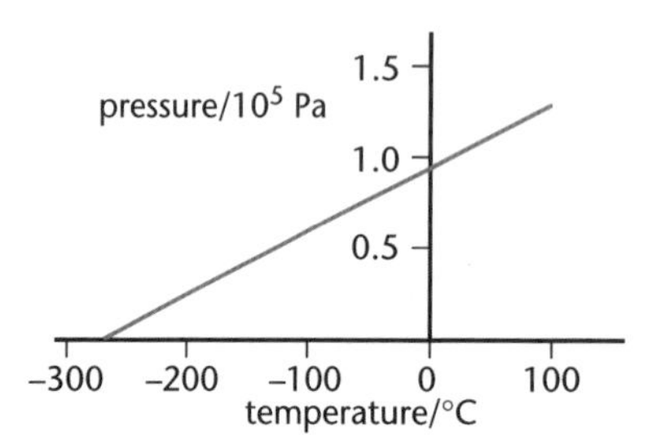

Fig. 10.3 When pressure is zero, temperature is –273°C.

The absolute zero of temperature is –273°C. This is the temperature when all the particles stop moving.

The kelvin temperature scale

The kelvin temperature scale is a scale that starts at absolute zero – so there are no negative kelvin temperatures.

The unit is the **kelvin** and the symbol is **K**. One kelvin is the same size as a Celsius degree, so to convert to kelvin you just add 273.

Fig. 10.4 shows how the Celsius and kelvin scales compare.

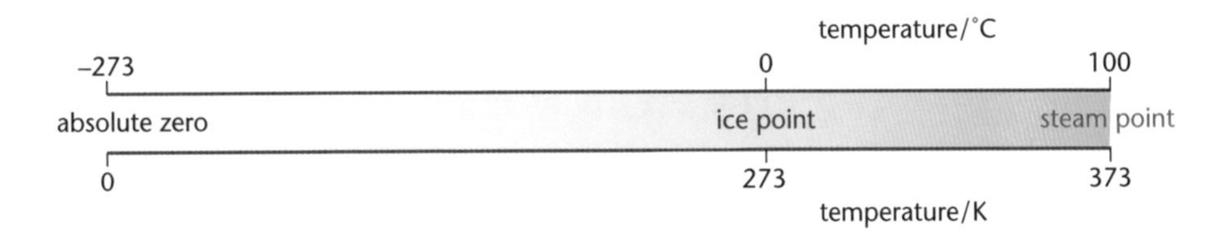

Note that the unit is K and not °K.

Fig. 10.4 Comparing the kelvin and Celsius temperature scales.

KEY POINT

To convert between temperature in kelvin and degrees Celsius:

temperature in K = temperature in °C + 273

temperature in °C = temperature in K – 273

The gas equations

EDEXCEL 360 P3.5

Temperature and energy

The advantage of using the kelvin temperature scale is that it is an absolute scale starting at zero. This means doubling the temperature of the gas doubles the average kinetic energy of the particles. (This is not true if the temperature is measured in °C.) Another way to say this is that the **temperature** of the gas is **directly proportional** to the **average kinetic energy** of its particles.

Pressure and temperature

For a mass of gas in a closed container with a constant volume, when the kelvin temperature is doubled the pressure doubles. (You can see in Fig.10.2 that this is not true if the temperature is in °C.) In other words, the **pressure** is **directly proportional** to the **temperature**.

KEY POINT

For a gas in a sealed container with constant volume:

$$\frac{\text{pressure}}{\text{temperature (in kelvin)}} = \text{constant}$$

$$\frac{P_1}{T_1} = \frac{P_2}{T_2}$$

The gas equation

This equation comes from collecting together the relationships for pressure temperature and volume.

KEY POINT

For a constant mass of gas, with pressure *P*, temperature *T* (in kelvin) and volume V:

$$\frac{P_1V_1}{T_1} = \frac{P_2V_2}{T_2}$$

10.2 Electrons and electron beams

Producing a beam

EDEXCEL 360 P3.5

The electrons are 'boiled off' from the hot metal filament.

The electrons in a metal filament (wire), are free to move, but are held inside the filament by the attractive forces between them and the positive ions (atom centres). Heating the metal filament gives the electrons more energy. If they have enough energy they can leave the filament in a process called **thermionic emission**.

Fig. 10.5 shows the structure of an electron gun, while Fig. 10.6 summarises how thermionic emission is used in an **electron gun** to produce a beam of electrons.

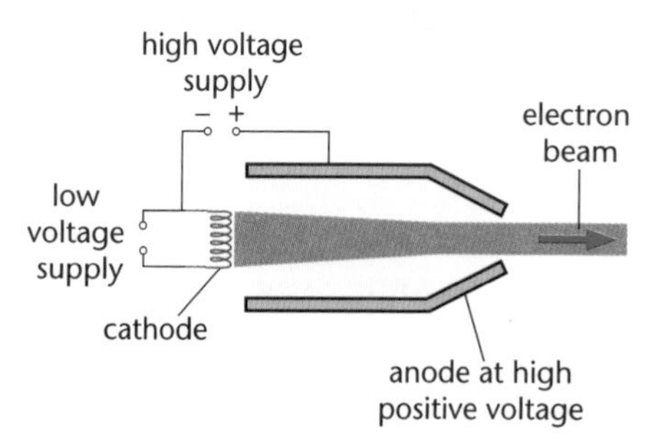

Fig. 10.5 The structure of an electron gun.

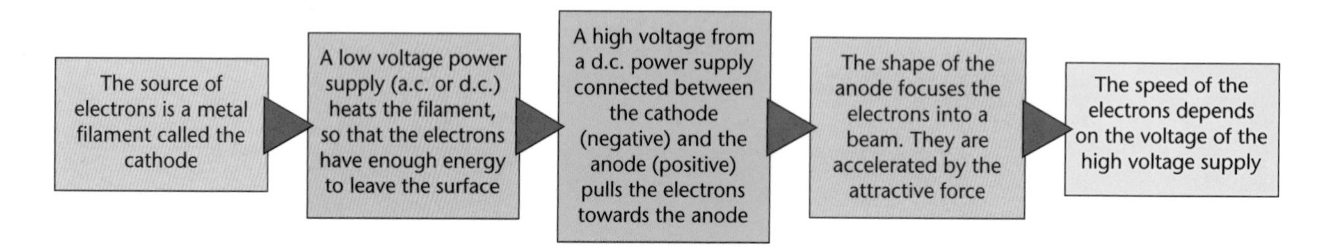

Fig. 10.6 How an electron gun produces an electron beam.

The kinetic energy of an electron in a beam depends on the size of the electric charge on the electron, (which is always e = 1.6×10^{-19} C, a value you will be given in the exam if you need it) and on the accelerating voltage.

KEY POINT

The kinetic energy of the electron = **charge on the electron** × **accelerating voltage**

$$KE = \frac{1}{2}mv^2 = eV$$

The beam of electrons is a flow of charge, so it is an electric current. Remember that because the electrons are negative the flow of positive charge (the current), is in the opposite direction to the flow of electrons. In other words the electrons go from the cathode towards the anode, but the current flows towards the cathode.

KEY POINT

current = number of electrons per second × charge on the electron

Deflecting the beam

An electron beam, or any beam of charged particles (for example ink drops), can be deflected by an electric field between two parallel charged metal plates. Fig. 10.7 a) shows how a beam of electrons is attracted to the positive plate, while b) shows a beam of heavier protons attracted to the negative plate. Because protons are heavier they are not deflected so much. The deflection increases if:

- there is a higher voltage on the parallel plates
- the charge on the particle is greater
- the mass of the particle is less.

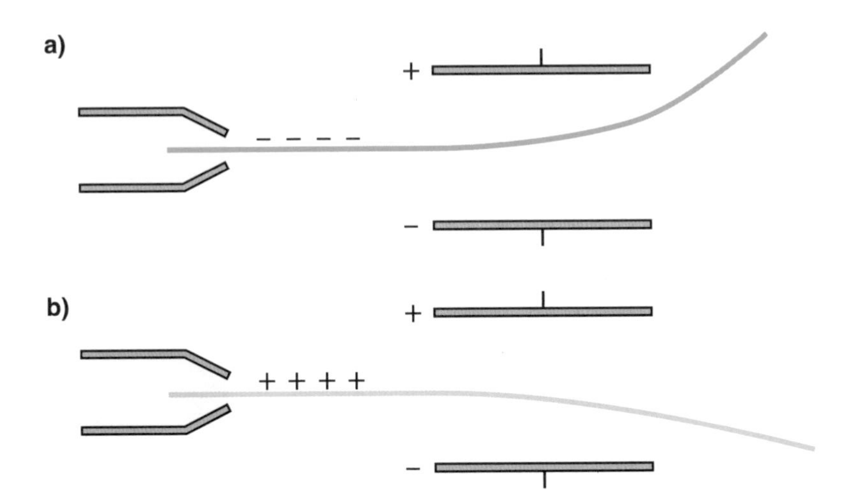

Fig. 10.7 A beam of charged particles is deflected by charged metal plates.

Uses of electron beams

EDEXCEL 360 P3.5

Flat screen TVs and computer monitors use different technology, as do digital oscilloscopes.

For many years, electron beams have been used in **TV picture tubes**, **computer monitors** and **oscilloscopes**. The electrons leave the electron gun and are steered by electric fields between metal plates (or, in a similar way, by magnetic fields) towards a fluorescent screen. Where the beam strikes the screen there is a bright area. The beam is scanned very quickly across the whole screen, making up a complete picture of light and dark spots.

X-ray tube

An **electron beam** is directed at a **tungsten metal target**. The electrons are decelerated and lose kinetic energy. Some of the energy heats up the metal target, and some energy leaves the metal as **X-rays**.

10.3 Atoms and nuclei

Radioactive emissions

EDEXCEL 360 P3.5

The properties of alpha (α), beta (β⁻) and gamma emissions are covered in Chapter 1 and Chapter 9. Another type of emission is **positron** (β^+) emission. A positron has the same mass as an electron and an equal but opposite charge. A positron is the anti-particle of the electron - if an electron and a positron meet they **annihilate** one another, leaving just energy.

When nuclear fission of uranium-235 occurs there are a few neutrons released. This **neutron** radiation is difficult to detect because neutrons are not charged.

Stability of the nucleus

EDEXCEL 360 P3.5

The stability of a nucleus depends on the balance of **protons** and **neutrons** it has. When the number of neutrons (N) is plotted against the number of protons (Z) all the stable isotopes lie on the stability line, shown in red on Fig. 10.8. Any isotope that does not lie on the line is unstable and will be radioactive.

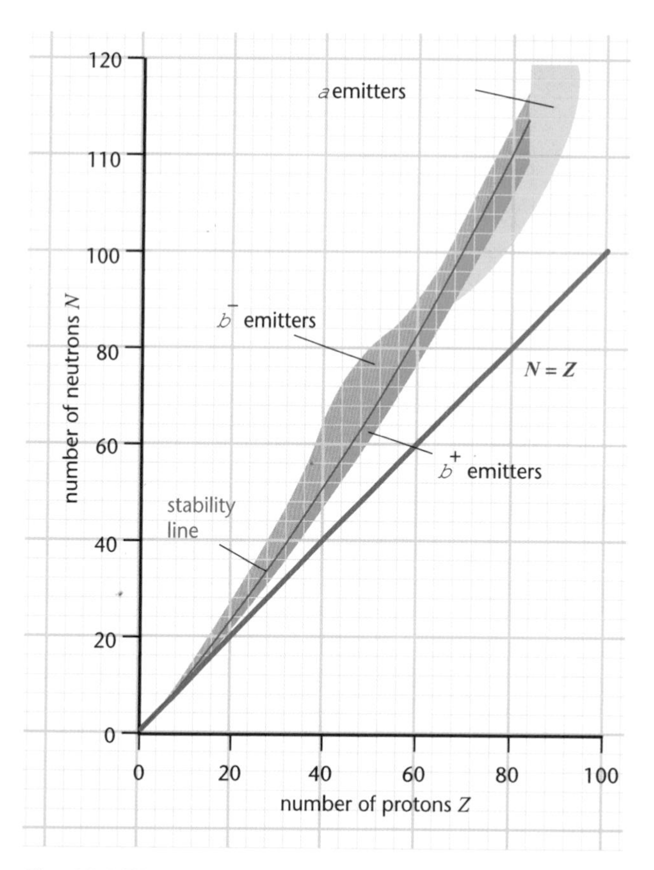

Fig. 10.8 The number of neutrons in a nucleus plotted against the number of protons.

Notice that small, stable nuclei have the same number of protons as neutrons, but as the nuclei get bigger they need extra neutrons to be stable. (The red line curves away from the blue N = Z line.) Above Z = 82 all the nuclei are in the green area, because all nuclei with more than 82 protons are unstable. They usually undergo alpha decay.

Three isotopes of carbon are carbon-11 carbon-12 and carbon-14.

- Carbon-12 is stable and lies on the red line.
- Carbon-14 is above the red line in the blue shaded section. It has too many neutrons to be stable and will decay by beta emission.
- Carbon-11 is below the line in the red shaded section. It has too few neutrons to be stable and will decay by positron emission.

β^- emission occurs when a neutron in the nucleus becomes a proton and an electron. β^+ emission occurs when a proton in the nucleus becomes a neutron and a positron. The nucleus often undergoes rearrangement after beta decay and gives out gamma radiation.

See Chapter 9 for more about nuclear equations.

KEY POINT

β^- decay: neutron → proton + electron

β^+ decay: proton → neutron + positron

The **positron** can be represented as $^{0}_{+1}e$. When it leaves the nucleus the proton number (atomic number) decreases by one and the mass number is unchanged.

10.4 Fundamental and other particles

Fundamental particles

EDEXCEL 360 P3.5

Fundamental particles are particles that are not made up from smaller particles, and this includes the **electron** and the **positron**. Other fundamental particles are **neutrinos**, **muons** and **tauons**. Each particle has a corresponding anti-particle. The anti-particle has the same mass and equal but opposite charge, (if it is charged.) Anti-particles are also called **anti-matter**. When a particle collides with its antiparticle the two annihilate each other and the energy is conserved as gamma rays. The reverse process also happens – gamma rays can turn into new particles and anti-particles.

The proton and neutron are not fundamental particles. They are made of quarks.

Scientists make new particles to study by colliding a beam of **positrons** and a beam of **electrons**. The resulting gamma rays then produce new particles – often ones that do not last very long. This is how **tauons** were discovered.

Quarks

EDEXCEL 360 P3.5

The **neutron** and the **proton** are each made of three particles called **quarks**. There are different types of quark. The proton and the neutron are made of two types – the **up quark** and the **down quark**.

KEY POINT

A proton is made of two up quarks and a down quark (uud).
A neutron is made of an up quark and two down quarks (udd).

Each quark has a **mass** of about **one third** the mass of the proton or neutron, so that the three together add to give the mass of the proton or neutron. (Mass and energy are interchangeable, so small differences in mass result from differences in the energy – some particles are more stable than others.) Each quark has electric charge, so that the total charge of the particle they make is a multiple of the charge on the electron (which has the symbol e).

Fig. 10.9 shows the particles that make up the proton and the neutron.

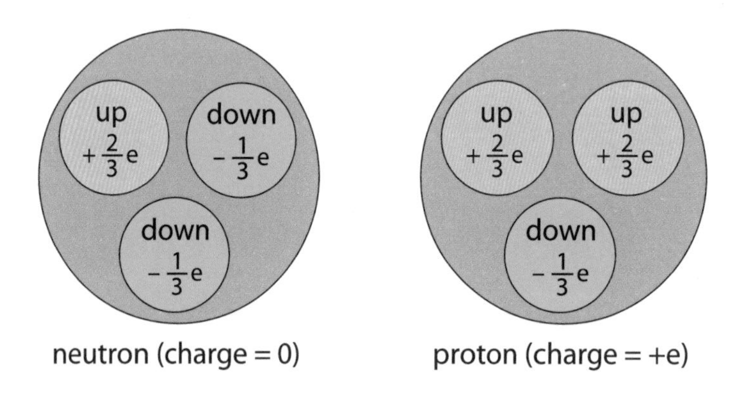

Fig. 10.9 The proton and the neutron are each made of three quarks.

When β^- decay occurs a down quark changes into an up quark. The neutron becomes a proton and an electron is emitted.

When β^+ decay occurs an up quark changes into a down quark. The proton becomes a neutron and a positron is emitted.

HOW SCIENCE WORKS

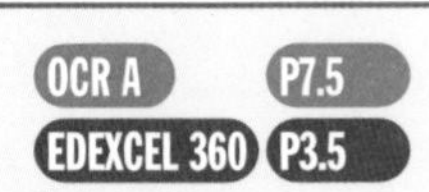

CERN

Research in the field of particle physics is very expensive. To produce beams of high energy particles requires large particle accelerators. In 1954, the European Particle Physics Laboratory was established near the Swiss city of Geneva. Instead of each country building its own particle accelerator, money from several European states was pooled and a large facility was built that all the scientists could use. There are now 20 European states that are members, but the 6500 visiting scientists represent 80 nationalities. The letters CERN stand for *Conseil Européen pour la Recherché Nucléaire*. The large electron positron collider (LEP) started operating in 1989. It was the largest scientific instrument ever built – the circular underground tunnel that the particles travelled along is 27 km long and passes under the border between Switzerland and France. The newest machine – the large hadron collider – occupies the same tunnel and started operating in 2007.

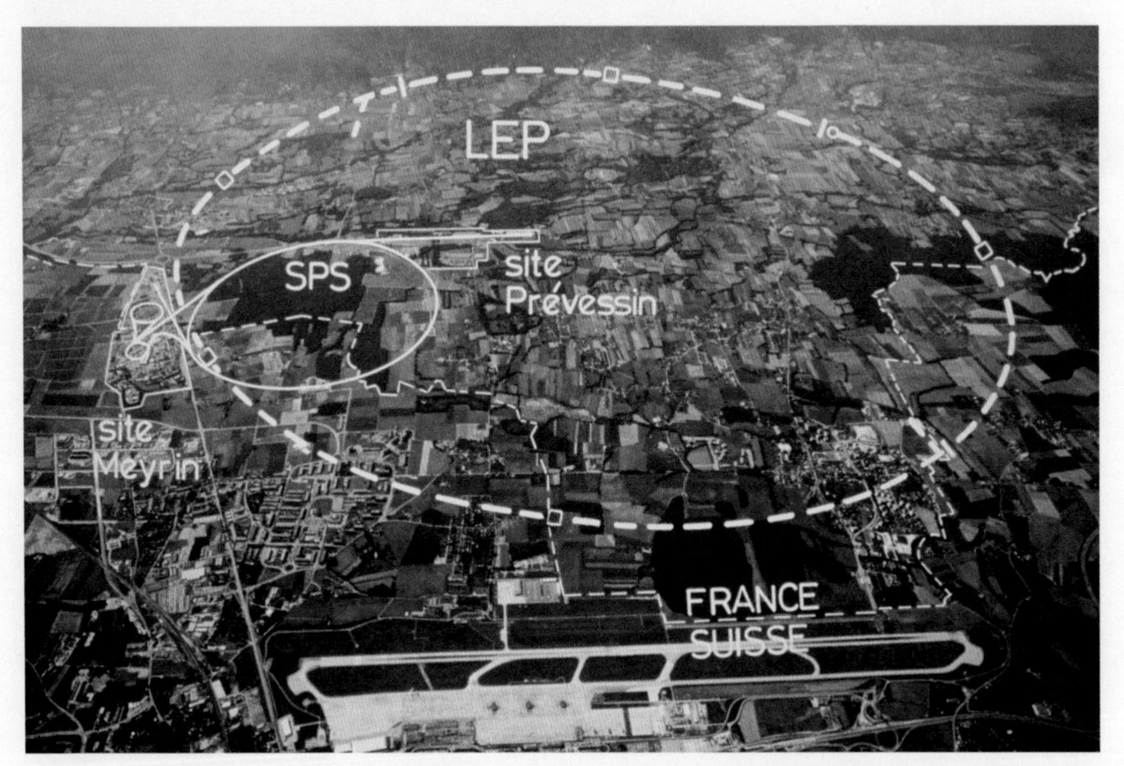

Scientists prepare for their experiments in their home laboratories. They go to CERN and carry them out, and they send the data back to their home laboratory for analysis. They have opportunities to discuss work and share ideas with other scientists. They collaborate on projects – a team making a new type of detector may work with a team trying to detect a new particle. These advantages often produce results more quickly than scientists working alone. For non-scientists projects like CERN provide jobs in the area – everything you can think of including, for example, shop assistants, taxi-drivers and doctors. Three thousand people work at CERN. Some disadvantages are that it may make living costs higher for local people, and, although funding for CERN is unlikely to stop, sometimes science facilities are closed or re-located, leaving the local community with few jobs.

HOW SCIENCE WORKS

The World Wide Web

Sometimes the spin-offs from scientific research can be quite unexpected. In 1989 Tim Berners-Lee, a physics graduate from Oxford University, was working as a software consultant at CERN. He had the idea for a project called the World Wide Web. It was designed to allow people to work together by combining their knowledge in a web of documents. The program was first made available within CERN in December 1990, and on the Internet in 1991. Scientists made it available, and now everyone uses it. For some groups it has huge advantages (examples are access to information, communication between people with similar interests, and banking and shopping) but there are some drawbacks (examples are the criminal activities that make use of the system like sharing child pornography, downloading films, music and books that have copyright without paying, and hacking into other computers). The challenge is to maximise the advantages and reduce the disadvantages as much as possible.

1. Describe two advantages of scientists collaborating on large projects. **[2]**
2. Describe an advantage and a disadvantage for a local community of having a large science facility close by. **[2]**
3. Give an example of
   (a) a group of people who benefit from the internet
   (b) a group for whom the internet has an undesirable effect. **[2]**
4. When deciding whether to make use of technology the right decision may be the one that gives the best outcome to the majority of people. Do you think it was the right decision to make the internet available to everyone? Explain your answer. **[2]**

Exam practice questions

1. Liquid nitrogen condenses at –196°C in kelvin this temperature is

A –469 K
B –77 K
C 41 K
D 77 K **[1]**

2. A phosphorus-32 nucleus decays to a sulphur nucleus by emitting a β^- particle. Which of these equations correctly shows the decay equation

A ${}^{32}_{15}P \rightarrow {}^{32}_{16}S + {}^{0}_{-1}e$
B ${}^{32}_{15}P \rightarrow {}^{33}_{15}S + {}^{1}_{0}e$
C ${}^{32}_{15}P \rightarrow {}^{32}_{14}S + {}^{0}_{-1}e$
D ${}^{32}_{15}P \rightarrow {}^{33}_{16}S + {}^{1}_{+1}e$ **[1]**

3. Which of the following groups of particles are all fundamental particles?

A neutron, proton, electron
B alpha particle, electron, quark
C electron, positron, neutrino
D electron, positron, neutron **[1]**

4. Write **T** by the **True** statements and **F** by the **False** statements
When a gas is compressed to half its volume:

(a) Its pressure doubles if the temperature is unchanged.
(b) Its temperature (in °C) doubles if the pressure is unchanged.
(c) The pressure and the temperature (in K) double.
(d) The average kinetic energy of the particles increases.
(e) The pressure increases because the particles make more collisions with each other. **[3]**

5. Match the situation shown in the left-hand boxes to the temperature in kelvin shown in the right-hand boxes.

Situation	**Temperature**
Body temperature 37°C	273 K
Absolute zero of temperature	0 K
Freezing temperature of water	236 K
November in the Antarctic –37°C	310 K

[3]

Exam practice questions

6. The stability of a nucleus depends on the neutron-proton ratio. Look at the graph shown in Fig. 10.8 (page 156).

(a) Explain how the number of neutrons compared to the number of protons in a stable nucleus changes as the proton number increases. **[2]**

(b) There are no nuclei that decay by alpha particle emission with less than 10 protons in the nucleus. Suggest a reason for this. **[1]**

7. An electron gun is used to produce a beam of electrons.

(a) What is the name of the process that results in electrons being released from the metal filament that forms the cathode? **[1]**

(b) Explain why the filament must be hot to release electrons. **[2]**

(c) The electrons are accelerated by a 4.2 kV voltage between the anode and cathode. Work out the kinetic energy in joules of an electron in the beam (charge on an electron $e = 1.6 \times 10^{-19}$ C). **[2]**

8. In the LEP collider at CERN a beam of positrons and a beam of electrons are collided.

The charge on the electron is $e = -1.6 \times 10^{-19}$ C
The mass of the electron is $m_e = 9.1 \times 10^{-31}$ kg

(a) What is **(i)** the charge on the positron?
(ii) the mass of the positron? **[2]**

(b) Explain what happens when an electron and a positron collide **[2]**

Gamma rays can produce new particles like tauons and muons. These are fundamental particles.

(c) Explain what is meant by a fundamental particle. **[1]**

The LEP has been replaced by a large hadron collider that will collide beams of protons.

(d) Complete this diagram of a proton with the name of the third particle and its charge.

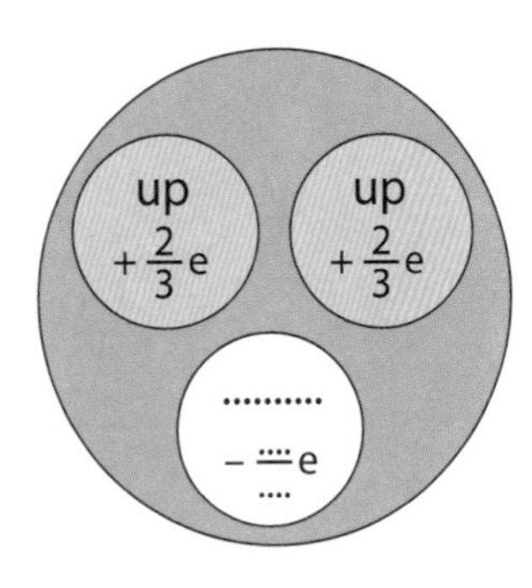

[2]

A beam of protons is accelerated through a voltage of 2500 kV.

(e) Calculate the energy of one proton. **[2]**

(f) Explain an advantage of scientists collaborating on a large project like CERN. **[1]**

More forces and motion

The following topics are covered in this chapter:

- *Movement in one and two dimensions*
- *Gravity, circular motion and orbits*
- *Turning effects*

11.1 Movement in one and two dimensions

Vectors and scalars

Vectors have **magnitude** and **direction**. Velocity, acceleration and momentum are examples of vectors.

Scalars have **magnitude** only. Speed and energy are examples of scalars.

A vector distance – distance in a certain direction – is usually given the symbol s.

When vectors are added direction must be taken into account. Section 6.2 shows how parallel forces can be added, taking direction into account, to give a **resultant force**.

Fig. 11.1 shows how to find the **resultant force** when two forces are acting at right angles on the same object.

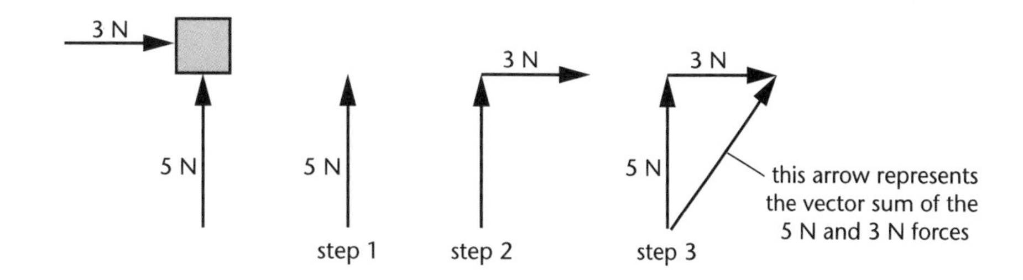

Fig. 11.1 The resultant of two forces at right angles.

Equations of motion

When the acceleration of an object is constant or zero, there are a number of equations you can use to work out acceleration, velocity, distance, or time.

KEY POINT

a = acceleration
u = intial velocity
t = time
v = final velocity
s = vector distance

$v = u + at$

$s = \frac{(u + v)}{2} t$

$v^2 = u^2 + 2as$

$s = ut + \frac{1}{2}at^2$

Example 1
A stone falls from the top of a cliff. How far will it fall before reaching a speed of 30 m/s? (g = 10 m/s^2)

1st step – choose which direction is positive, and write down the variables you know, and the one you want to work out:

down is + u = 0 m/s v = 30 m/s s? a = 10 m/s^2

Choose the best equation: $v^2 = u^2 + 2as$

$$(30\text{ m/s})^2 = 0 + 2 \times (10\text{ m/s}^2)\ s$$

$$s = 900 \div 20 = 45\text{ m}$$

Example 2
A stationary rollercoaster accelerates for 3 seconds. It has travelled 40.5 m. What is its acceleration?

Forward is + u = 0 m/s t = 3 s s = 40.5 m a?

$$s = ut + \frac{1}{2}at^2$$

$$40.5\text{ m} = 0 + \frac{1}{2}a(3s)^2$$

$$a = \frac{2 \times 40.5}{9} = 9\text{ m/s}^2$$

Projectiles

A **projectile** is the name used for an object that is thrown or fired horizontally in the Earth's gravitational field, and for which air resistance is negligible. Examples are balls and bullets. There is no horizontal force on the projectile once it is in flight. The only force is the downward force of gravity. This means a projectile:

- has constant horizontal velocity (horizontal acceleration = 0)
- is accelerating downwards with constant acceleration due to gravity (vertical acceleration = g = 10 m/s^2 downwards).

Fig. 11.1 shows that two forces can be considered separately to find the resultant force. In the same way the horizontal motion of a projectile can be considered separately to the vertical motion to work out, for example, the distance travelled, or the time the projectile is in the air.

The resultant velocity of the projectile at any time is the vector sum of the horizontal and the vertical velocities.

distance travelled in equal time intervals

Remember that the time the projectile takes to fall vertically to the ground is the same as the total time it moves for in the horizontal direction.

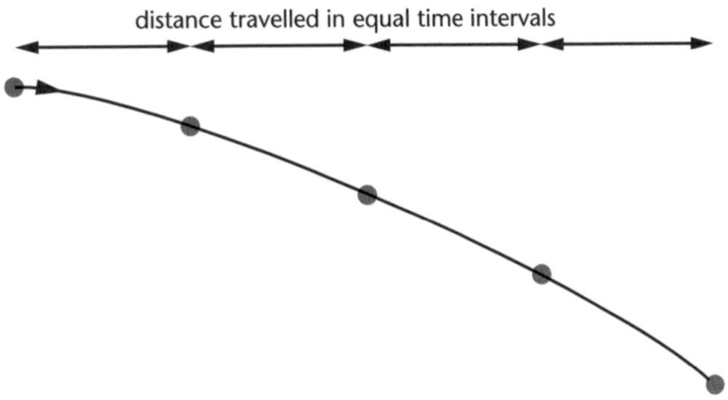

Fig. 11.2 A projectile moves with constant horizontal velocity while accelerating vertically.

Example

A ball is thrown at 12 m/s horizontally, from a point 1.25 m above the ground. How far does it travel?

($g = 10\ \mathrm{m/s^2}$)

It will travel until it hits the ground so you need to work out the time this takes:

Vertically: (down = +) $u = 0$ $s = 1.25\ \mathrm{m}$ $a = 10\ \mathrm{m/s^2}$ $t?$

$$s = ut + \frac{1}{2}at^2$$

$$1.25 = 0 + \frac{1}{2} \times 10\,t^2$$

$$t = \sqrt{0.25}\ \mathrm{s} = 0.5\ \mathrm{s}$$

Horizontally: (forward = +) $u = 12\ \mathrm{m/s}$ $s = ?$ $a = 0$ $t = 0.5\ \mathrm{s}$

$$s = ut + \frac{1}{2}at^2$$

$$s = (12 \times 0.5 + 0)\ \mathrm{m} = 6\ \mathrm{m}$$

At any point in its trajectory (path) the velocity of the projectile is the vector sum of the horizontal and vertical velocities, as shown in Fig. 11.3.

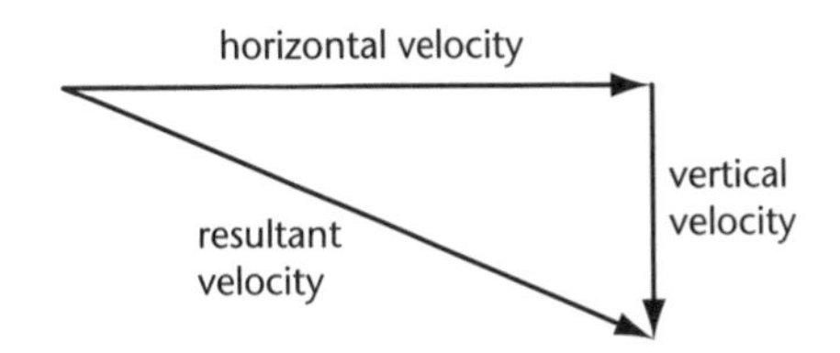

Fig. 11.3 The velocity is the vector sum of the horizontal and vertical velocities.

Action and reaction forces

This section builds on section 6.2 Interaction pairs – action and reaction forces. Fig. 11.4 shows the opposite **reaction** force to the weight of a man standing on the Earth. The reaction force attracts the Earth upward towards the man with a force equal and opposite to his weight. So why don't we notice the Earth being attracted upwards? The answer is that because the mass is so large, the acceleration is very tiny ($F = ma$.).

the Earth pulls the person

the person pulls the Earth

Newton's third law: pairs of forces that are equal in size and opposite in direction.

Fig. 11.4 An action and reaction pair of forces.

Conservation of momentum

Conservation of momentum can be used to explain the behaviour of objects that collide or explode apart. Before an **explosion** of a stationary object the momentum is zero. After the explosion, conservation of momentum tells us that the vector sum of the momentum of all the separate pieces will still be zero.

Momentum and collisions are covered in section 6.3.

Rocket propulsion

A rocket is designed so that when the fuel and the oxygen combines explosively, the exhaust gases can only leave in one direction. As the exhaust gases all move in one direction there is an equal but opposite force on the rocket in the other direction, as shown in Fig. 11.5. The momentum of the exhaust gases is equal and opposite to the momentum of the rocket.

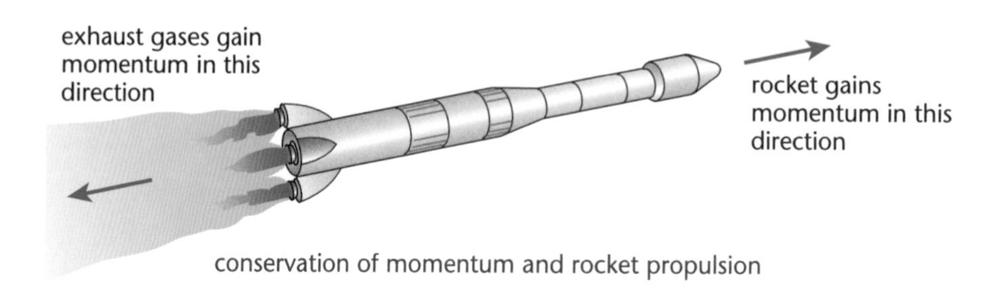

Fig. 11.5 Conservation of momentum and rocket propulsion.

Recoil

In films, you often see large cannon recoil as the cannon ball leaves the barrel. The momentum of the cannon is equal and opposite to the momentum of the cannon ball. (Modern guns are sometimes designed so this momentum is absorbed internally, by moving parts inside the gun.)

Example

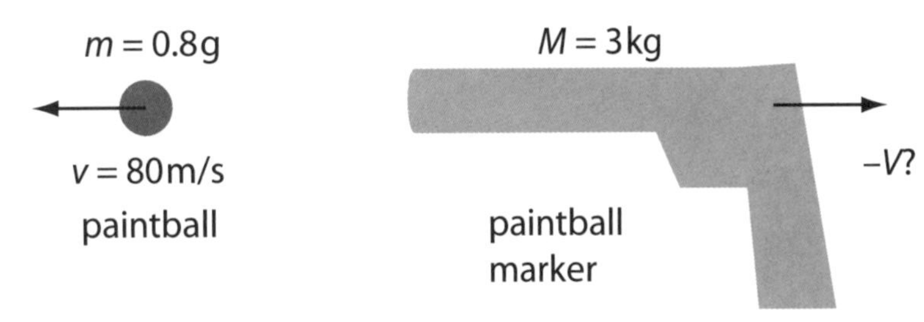

Fig. 11.6 Firing a paintball.

When a 0.8 g paintball is fired at 80 m/s from a 3 kg paintball marker, what is the recoil velocity of the marker?

Paintball: mass, m = 0.8 g velocity v = 80 m/s

Marker: mass, M = 3 kg = 3000 g, velocity, V = ?

Conservation of momentum: $m\,v = M\,V$

$$V = \frac{0.8 \times 80}{3000} = 0.01 \text{ m/s}$$

11.3 Turning effects

Clockwise and anticlockwise moments

AQA P3.13.1

When a force acts on an object that is free to rotate about a **pivot** (a fixed point), the force has a turning effect, which is called a **moment**.

If the line of action of the force passes through the pivot, there is no turning effect.

The size of the **moment** depends on:

- the size of the force
- the distance of the point where it is applied from the pivot
- the angle at which the force is applied.

KEY POINT

The moment (or turning effect) of a force is worked out from:

moment = force × perpendicular distance to the pivot

The moment of a force is measured in Nm.

The moment can be clockwise or anticlockwise.

KEY POINT

The principle of moments

When a system is balanced, about any axis:

The sum of the anticlockwise moments = the sum of the clockwise moments.

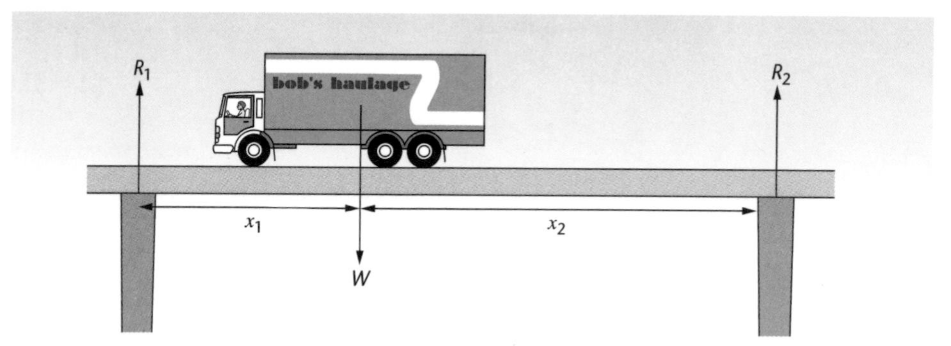

Fig. 11.7 The clockwise moments about any axis on the bridge equal the anticlockwise moments.

Taking moments about an axis through the left-hand support:

$$R_2 \times (x_1 + x_2) = W \times x_1$$

Or, taking moments about an axis through the right-hand support:

$$R_1 \times (x_1 + x_2) = W \times x_2$$

Notice that as the lorry moves across the bridge, x_1 gets smaller and x_2 gets bigger. At the same time, R_1 gets bigger and R_2 gets smaller so that the equations above are still true.

Centre of mass

AQA P3.13.1

The **centre of mass** of an object is the point at which all the mass of an object seems to be concentrated. It is the point where the weight appears to act. For a human body it is just behind the 'tummy button'. You can see from Fig. 11.8 the point does not have to be inside the object – for a metal ring it is in the centre of the ring.

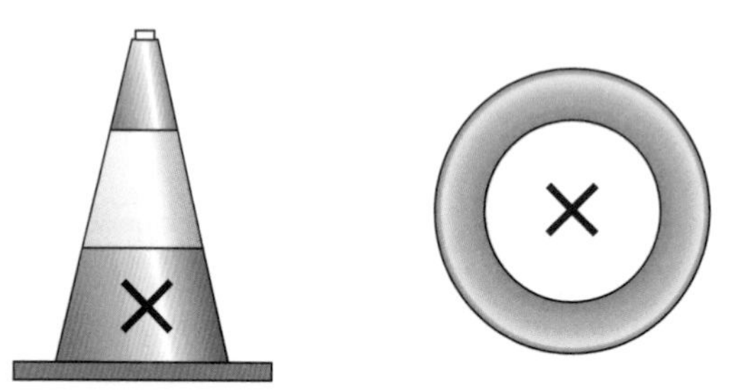

Fig. 11.8 The centre of mass is the point where weight seems to act.

If in object is suspended freely it will stop swinging and hang so that the centre of mass is directly below the point of suspension, as shown in Fig. 11.9.

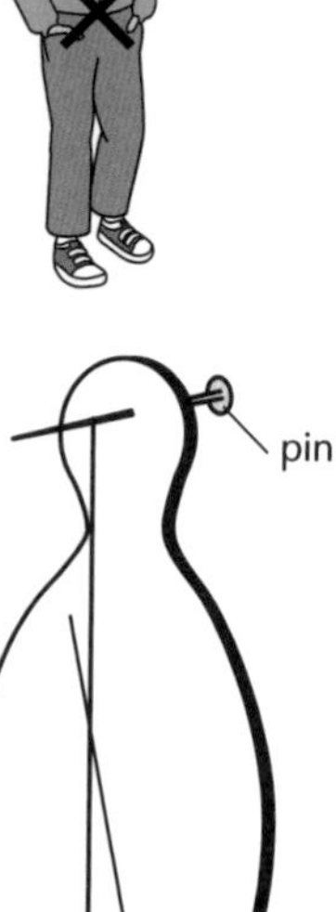

Fig. 11.9 Finding the centre of mass of a piece of card.

The centre of mass of a thin sheet of material can be found by:

- suspending the shape from a pin in a loose-fitting whole so that it is free to swing
- using a plumb-line to mark the vertical line from the pinhole
- repeating this at other points.

All the lines will cross at the same point, and this point is the **centre of mass**. You can check this by turning the shape to a horizontal position and balancing it on your finger. The only place it will balance is when your finger is at the centre of mass.

For a symmetrical object, the centre of mass is on the axis of symmetry.

Stability

P3.13.1

If the centre of mass of an object moves so that the line of action of the weight is outside the base of the object, there will be a moment on the body that will make it topple over. Fig. 11.10 shows that the **stable** traffic cone has a moment that will make it fall back onto its base, when it is allowed to move, whereas the **unstable** bowling pin has a moment that will tip it over.

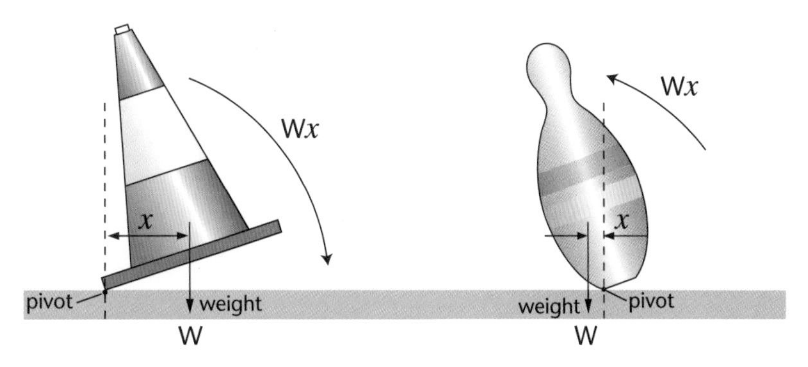

Fig. 11.10 The traffic cone will not topple at this angle, but the bowling pin will topple.

The traffic cone is more **stable** than the bowling pin because:

- the traffic cone has a wider base
- the traffic cone has a lower centre of mass.

An object may have a low **centre of mass** because of its shape, like the traffic cone, or because there is more mass near the base. Double-decker buses are more stable than they look because the extra mass of the engine and transmission near the ground gives them a low centre of mass.

11.3 Gravity, circular motion and orbits

Circular motion

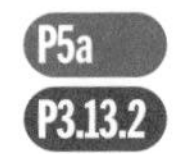

Acceleration is change of velocity. Velocity is speed and direction.

When an object moves in a circular path it is constantly changing **direction**. Its velocity is changing so it is **accelerating**, even if its speed is constant.

There must be a resultant force on the object to make it accelerate (change direction). This force is called a centripetal force, it changes the direction of the object not its speed.

For an object to move in a circle there must be a resultant force on it towards the centre of the circle. This force is called the centripetal force.

Fig. 11.11 shows an object on a string moving in a circle. The **centripetal force** is provided by the **tension** in the string. Cut the string and the object flies off at a **tangent** to the circle – a straight line.

Fig. 11.11 A circular path needs a resultant force towards the centre of the circle.

When a car goes round a circular track, the **centripetal force** is provided by the friction between the tyres and the road.

For astronomical objects that are in orbit, the **centripetal force** is provided by **gravity**, for example, planets orbiting the Sun (or the Moon and artificial satellites orbiting the Earth).

To make an object move in a circle, the **centripetal force** increases as

- the mass of the object increases
- the speed of the object increases
- the radius of the circle decreases.

More about orbits

This section follows from section 5.3 Gravity.

Gravity is a force of attraction between objects with mass.

The gravitational force:

- increases with mass (if either of the masses is doubled, the force is doubled)
- decreases with distance (if the distance between two objects is doubled, the force of gravity between them is a quarter the original size).

To stay in orbit at a particular distance, smaller objects, including planets and satellites must move at a particular speed around larger objects, so that the centripetal force that is needed is equal to the gravitational force. Objects like comets, that change their distance from the Sun, speed up as they get close to the Sun and slow down as the get further away. The further away an object is, the longer it takes to make a complete orbit.

For AQA you need to know that the orbit of a planet is an ellipse, with the Sun at one focus. To understand what the focus of an ellipse is, look at Fig. 11.12. This shows how to draw an ellipse with a loop of string, 2 pins and a pencil. The string must be kept taut as the pencil is moved around.

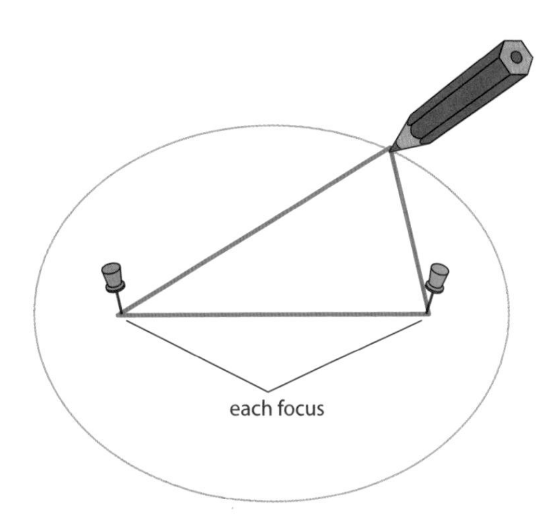

Fig. 11.12 How to draw an ellipse.

Satellite orbits

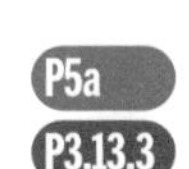

OCR B P5a
AQA P3.13.3

Satellites in **low Earth orbits** orbit the Earth in a few hours. The higher a satellite is above the surface of the Earth, the slower its speed, because the gravitational force is less. At a particular height above the equator of the Earth, the time to complete one orbit (called the **period**) will be exactly 24 hours. This is called the **geostationary orbit** because a satellite placed at this height will hover above the same place on the Earth as the Earth rotates and appear to be stationary.

Uses of satellites

Low polar orbits are orbits that go over both the poles. Each time the satellite completes an orbit the Earth has rotated a bit further, so over 24 hours the satellite can monitor the whole surface of the Earth. This can be useful for:

- weather satellites (monitoring large areas of the Earth)
- imaging the Earth's surface.

A satellite in geostationary orbit can be used to relay signals from one point on Earth to another. This is useful for:

- communications (for example, satellite TV and radio)
- weather forecasting (for detailed coverage of one area).

It is more expensive to launch a geostationary satellite, so sometimes several low orbit satellites are used for communication, rather than one geostationary satellite.

HOW SCIENCE WORKS

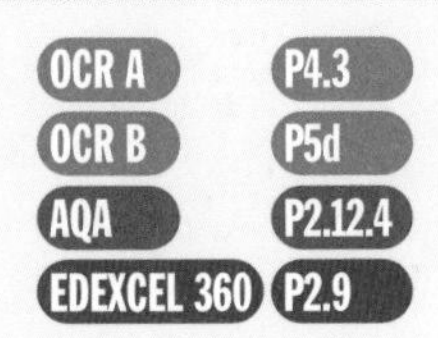

Seatbelts – Who benefits?

In a collision, if they are not wearing seatbelts, the driver and passengers continue moving forward with the same momentum until they hit something to stop them. This could be the windscreen, or the road if they are thrown out of the vehicle. For rear seat passengers, it could be the seats in front. All of these surfaces are hard and stop the body very quickly. The forces involved in such a sudden change of momentum, (see table 1) are often fatal. So seatbelts are 'A Good Thing' – they save lives.

Conditions: Seatbelt/speed	Distance over which body stops (m)	Time the body takes to stop (s)	Momentum of 80 kg body (kg m/s)	Stopping force on body (kN)
Seatbelt at 30 mph (13 m/s)	0.45 m	0.069	1040	15
No seatbelt at 30 mph (13 m/s)	0.06 m	0.009	1040	116
Seatbelt at 70 mph (31 m/s)	0.45 m	0.029	2480	86
No seatbelt at 70 mph (31 m/s)	0.06 m	0.004	2480	620

Table 1. How the force on the body varies depending on whether a person (mass 80 kg) is wearing a seatbelt, at 30 mph (13 m/s) and at 70 mph (31 m/s)

The statistics speak for themselves, the government estimates that since seat belts became compulsory in 1983, front seatbelts have reduced deaths by 370 and serious injuries by 7000 per year. (For rear seat passengers the reductions are 70 deaths and 1000 serious injuries per year.)

So is it all good news? A study of 19 000 cyclist and 72 000 pedestrian casualties from the time when compulsory seatbelts were introduced, suggests a correlation between the number of drivers wearing their seatbelts and the number of injuries to pedestrians and cyclists. Seatbelt wearing drivers were 11–13% more likely to injure pedestrians and 7–8% more likely to injure cyclists. Can this be coincidence or does wearing seatbelts cause more pedestrian deaths, and more cyclist deaths, because drivers feel safer and drive faster?

Table 2 shows how the chances of surviving a fatal collision reduce as speed increases.

Car speed (mph)	Chance of survival
20	95 %
30	80%
40	20%

Table 2 Chance of child pedestrian surviving a collision with a car.

Of course, supporters of seatbelts say that more people are saved by seatbelts than are killed by drivers speeding, but that is not much comfort to the family of a dead victim. Ways of reducing vehicle speeds include fines for drivers caught by speed cameras and traffic calming humps. No one is saying don't wear a seatbelt – just consider other road users and stick to the speed limits, (or a lower speed when necessary).

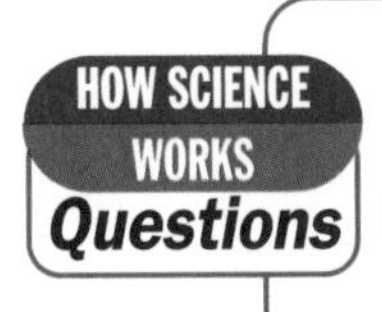

1. The author suggests that drivers wearing seatbelts causes more pedestrian and cyclist deaths. Is this true? Explain your answer. **[1]**
2. Use the information that explains why the people in a car are more seriously injured if they don't wear a seatbelt to help you write a scientific explanation of why pedestrians are more seriously injured at greater speeds. **[3]**
3. Suggest why drivers drive faster if they are wearing a seatbelt. **[2]**
4. According to the article:
   (a) Which groups benefit from the seatbelt law?
   (b) Which groups do not benefit? **[4]**
5. Suggest an argument that says it was right to pass the law even if some pedestrians died as a result. **[1]**
6. The article suggests that the development of faster and safer cars by scientists has led to more death and injury for pedestrians. What developments does the article mention that may help to reduce this unintended impact? **[3]**

Exam practice questions

1. A squash ball is hit and starts moving horizontally at 40 m/s. When it hits the wall it has dropped 0.2 m from its height when hit. How far away is the wall?

A 2 m
B 4 m
C 8 m
D 16 m **[1]**

2.

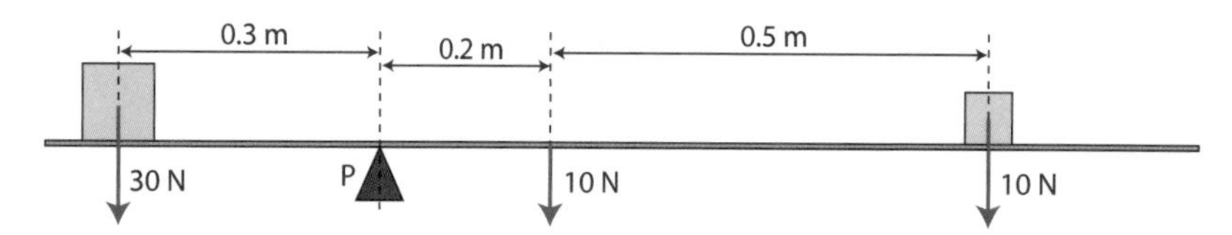

Fig. 11.13

This diagram shows a seesaw pivoted at point P. The seesaw weighs 10N as shown. The seesaw:

A will turn anticlockwise
B is balanced
C will turn clockwise
D you can't tell **[1]**

3. A satellite is placed in geostationary orbit. This means:

A It passes overhead once every 24 hours
B There is a force keeping it in orbit above the same point on the Earth's surface
C There is no force of gravity acting on it
D It will be moving more quickly than a satellite in a low Earth orbit **[1]**

4. Write **T** for the **True** statements and **F** for the **False** statements.

(a) Force is equal to rate of change of momentum.
(b) When objects explode the total momentum of the pieces is zero.
(c) When two objects collide the force on the more massive object is smaller than the force on the object with less mass.
(d) Conservation of momentum says that the momentum of an object does not change when a force acts on it.
(e) Rocket propulsion works because the momentum of the exhaust gases is equal and opposite to the momentum of the rocket.
(f) The momentum of a 1000 kg car travelling at 30 m/s is 30 000 kg m/s. **[3]**

5. Draw a straight line to join the motion with the correct condition.

Motion	**Condition**
Circular	Momentum is equal and opposite
Projectile	Constant acceleration at right angles to direction of travel
Recoil	Acceleration in the vertical but not the horizontal direction

[3]

Exam practice questions

6. In a test, a 1200 kg car crashes into the back of a 3000 kg truck:

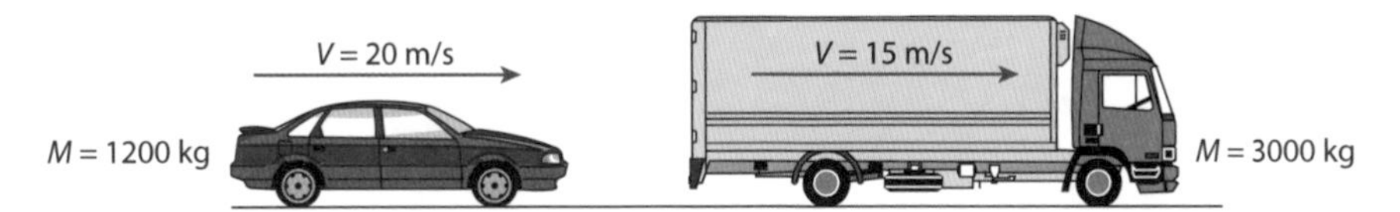

The car has a velocity of 20 m/s and the truck has a velocity of 15 m/s.

(a) Use the equation momentum = mass x velocity to calculate the momentum of **(i)** the car and **(ii)** the truck before the collision **[2]**

(b) What is the total momentum of the car and truck before the collision? **[1]**

(c) What will be the total momentum of the car and truck after the collision? **[1]**

(d) If the car and truck are stuck together after the collision, what will be the speed of the car and truck after the collision? **[2]**

7. An electron is accelerated and has momentum = 3.8×10^{-25} kg m/s.

The antiparticle of the electron is a positron, it has the same mass and opposite charge to the electron.

(a) What is the momentum of a positron travelling in the opposite direction towards the electron at the same speed as the electron?

(b) What is the total momentum of the two particles as they collide? **[2]**

8.

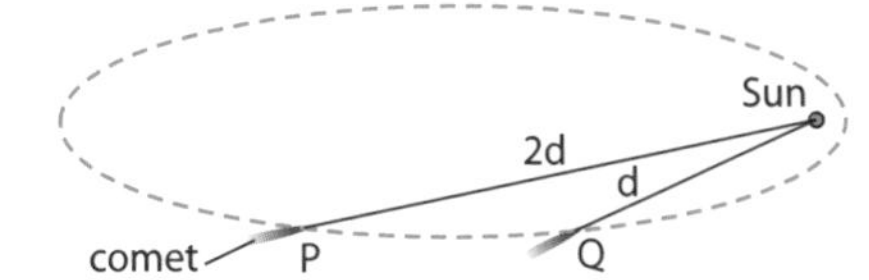

Fig. 11.15 A comet orbiting the Sun.

(a) What force keeps the comet in orbit around the Sun?

(b) In what direction is this force on the comet ?

(c) The force on the comet at P is ***F***. What is the force on the comet when it reaches point Q? Choose from: **0.25*F*** **0.5*F*** ***F*** **2*F*** **4*F***

(d) Explain whether the comet moves faster at P or at Q. **[5]**

9.

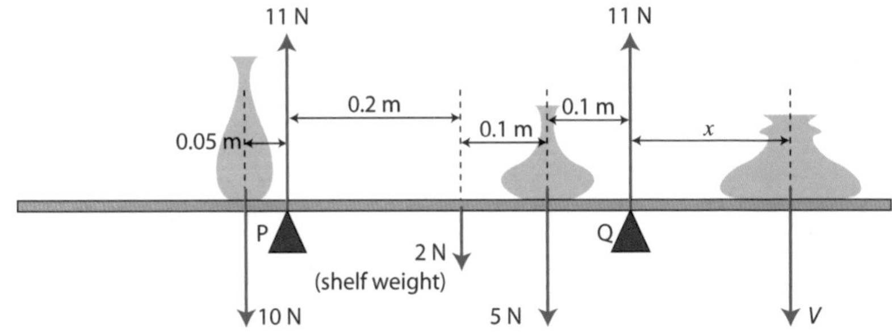

Fig. 11.16 shows a shelf weighing 2 N that is resting on 2 brackets P and Q. Three vases have been placed carefully on the shelf so it does not tip.

(a) By considering the upward and downward forces on the shelf calculate the weight V of vase 3. **[1]**

(b) By taking moments for all the forces about Q work out the distance x cm from bracket Q where the vase 3 will balance. **[2]**

(c) **(i)** Which of the three vases is the most stable design?
(ii) Give **two** reasons for answer. **[3]**

Chapter 12

Stars and the Universe

The following topics are covered in this chapter:

- *Observing the solar system and stars*
- *The distance to a star or galaxy*
- *The life of stars*
- *Lenses and telescopes*

12.1 Observing the Solar System and stars

The Solar System

OCR A 

P7.1

The Sun appears to travel from East to West across the sky once every **24 hours**. This is called a **solar day** and happens because the **Earth rotates**.

The Moon appears to travel from East to West across the sky more slowly, about once every **25 hours**. This is because the **Earth rotates** and the **Moon orbits** the Earth once every 28 days. Each time the Earth completes one rotation, the Moon has moved a bit further round its orbit, and it takes about another hour before it is in the same position again. This movement also results in the **phases of the Moon** (the changing appearance of the Moon over 28 days) as shown in Fig. 12.1.

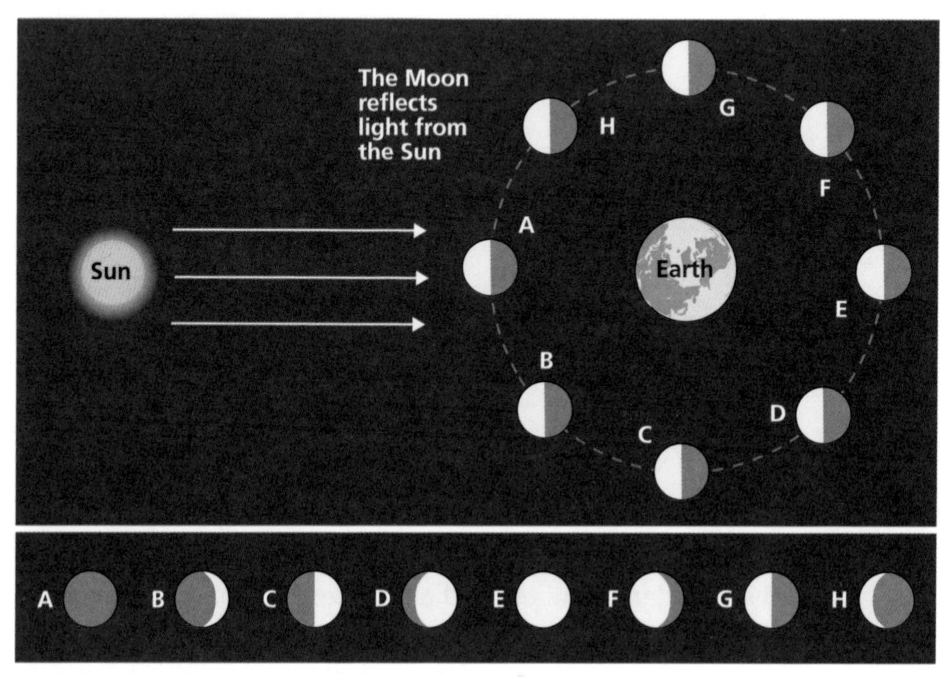

Fig. 12.1 The phases of the Moon (not to scale).

The sidereal day

The **stars** appear to travel from East to West across the sky once every **23 h 56 min**. This is called a **sidereal day**, it is 4 minutes less than a solar day. It happens because the **Earth rotates** and the **Earth orbits** the Sun. One complete rotation of the Earth through 360° is a sidereal day (23 h 56 min) but the Earth must rotate for an extra 4 minutes before the Sun is overhead again. This is shown in Fig. 12.2, where point X on the Earth is not in line with the Sun after rotation through 360°.

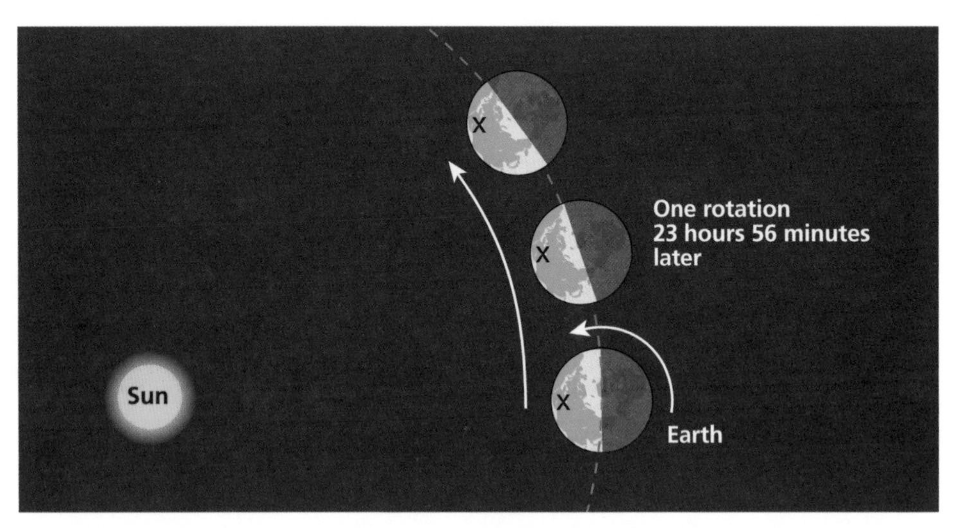

Fig. 12.2 The Earth must rotate more than 360° for the Sun to be overhead again.

To understand the movement of stars, imagine you have a black umbrella with white dots on the inside to represent stars. If you turn the handle so the umbrella rotates, as you look up the 'stars' (dots) move in circles. The Pole star, which does not move, is at the centre where the handle is joined to the umbrella. This is how the stars appear to move, but really the Earth is spinning and the stars are still (as if you are spinning and the umbrella is still). The handle is the axis of rotation of the Earth. The stars can only be seen from the part of the Earth facing away from the Sun, so over the year these change, as shown in Fig. 12.3. There is another complete set of stars seen from the southern hemisphere – one of the most well known constellations is the Southern Cross, which appears on the flags of Australia and New Zealand.

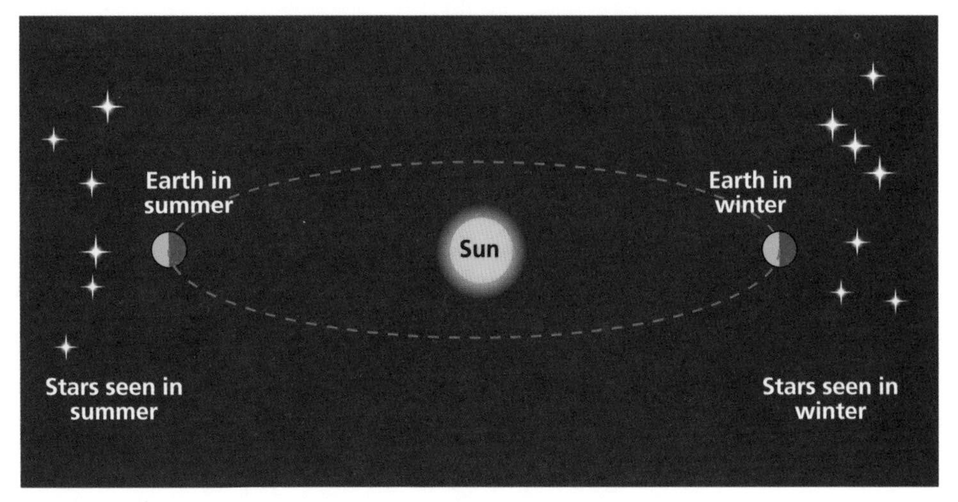

Fig. 12.3 Different stars are seen in summer and winter.

Eclipses

A solar eclipse occurs when the Moon gets between the Sun and the Earth and blocks the light. Although the Moon is small it is close to the Earth so can completely block the Sun, but only over a small part of the Earth's surface.

Eclipses are rare because the orbit of the Earth round the Sun and the orbit of the Moon around the Earth are tilted so the Moon often passes above or below the line between the Sun and the Earth. Fig. 12.4 shows the conditions for an eclipse.

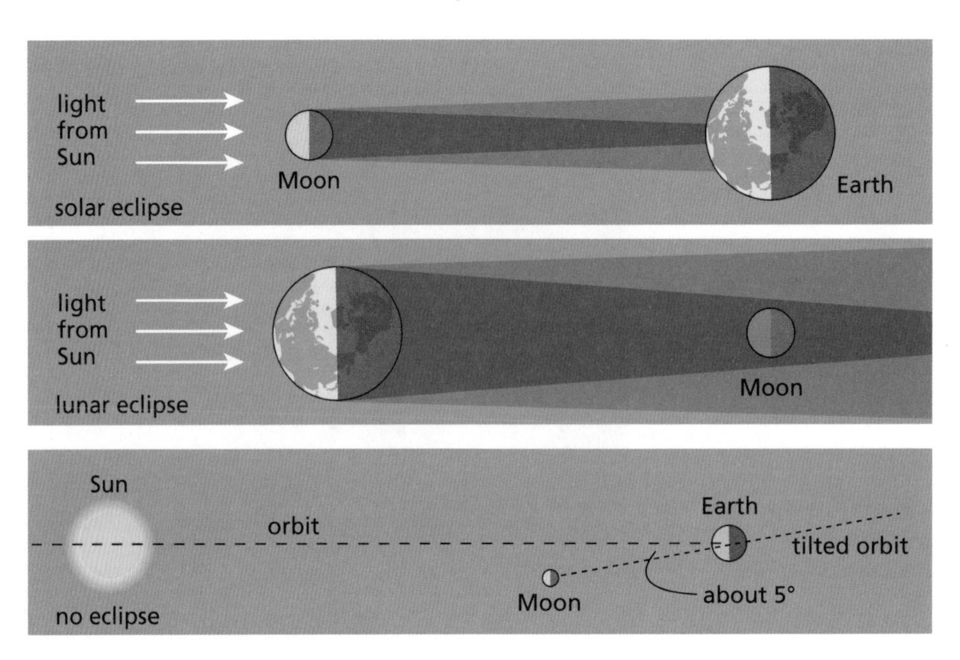

Fig. 12.4 Conditions for an eclipse (not to scale).

Planets

The planets that can be seen with the naked eye are Mercury, Venus, Mars, Jupiter and Saturn. These appear to move across the sky, like the stars, in a sidereal day, but they also change their positions relative to the stars in complicated patterns. This is because the **planets**, including the **Earth** are **orbiting the Sun**. Against the background of stars, a planet moves a little each day in one direction, and then sometimes it slows down and starts to move in the opposite direction. This is known as **retrograde motion**. Imagine a number of cars going around a circular track at different speeds and you are in one of the cars. As you look at the other cars sometimes they appear to be going forwards, and sometimes backwards, depending on which part of the track they are on, and your position.

Positions in the sky

The positions of stars and other astronomical objects in the sky are described by two angles. To give the position of a star:

- start by pointing North. Turn westwards until you are pointing at the horizon directly below the star. The angle you have turned through is the first angle
- move your arm upwards until you are pointing directly at the star. The angle between the horizontal and the star direction is the second angle.

12.2 The distance to a star or galaxy

Parallax

OCR A P7.3

The idea of parallax was introduced in section 5.5 Difficult observations. Fig. 12.5 shows the angle of parallax. The smaller the angle, the further away the star.

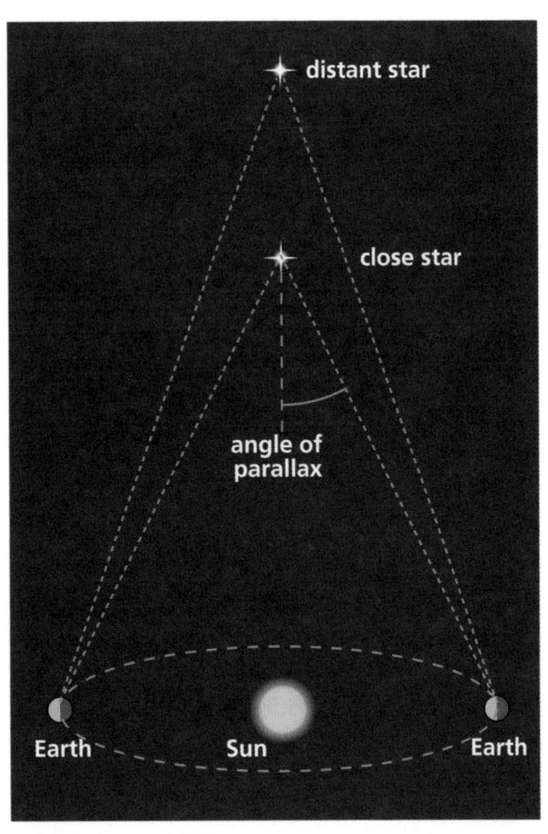

Fig. 12.5 The angle of parallax for a close and a distant star.

KEY POINT

The angle of parallax is half of the angle moved against a background of stars in six months.

The angles are very small – a sixtieth of a degree (1°) is a minute of arc (1′), and a sixtieth of a minute is a second of arc (1″). Angles of parallax are about a second of arc. Astronomers use a unit of distance called the **parsec**, which is based on the **angle of parallax**.

KEY POINT

A parsec (pc) is the distance to a star with a parallax angle of one second of arc.

A **parsec** (3.1×10^{16} m) is similar to a **light year** (9.5×10^{15} m) and **interstellar** distances (the distances between stars) are typically a few parsecs.

Intrinsic brightness and observed brightness

OCR A 

The **intrinsic brightness** of a star refers to the actual brightness the star. If you could place all stars at the same distance from Earth as the Sun you could compare their real (intrinsic) brightness. Two things affect intrinsic brightness:

- the **size** of the star, a larger surface radiates more energy each second. For example, red giants are bright because they are big and white dwarfs are dim because they are small
- the **temperature** of the star. A hotter surface radiates more energy each second.

The **observed brightness** is how bright the star appears to us on the Earth. This depends on:

- the intrinsic brightness
- how far away the star is. The further away the star the dimmer it will appear

If we know the intrinsic brightness of a star, we can work out how far away it is from the observed brightness. The problem is – how to find out the intrinsic brightness.

Temperature and colour

Objects radiate electromagnetic radiation which depends on the temperature. An object may appear black because it is not reflecting any light, but it is giving out infrared radiation. When heated, at about 600°C, it will glow red. At a higher temperature it glows yellow. It will still be radiating infrared and red light, but there will be yellow light as well. At an even higher temperature it will glow white.

Stars are different colours and the **colour** depends on the **temperature** of the star. The Sun is hot enough to glow white, its peak wavelength is in the yellow part of the spectrum. Some hotter stars glow blue-white.

The colour of a star depends on its temperature. Cool stars glow red. Hot stars glow blue-white.

Fig. 12.6 shows the intensity of the radiation at different frequencies, for a hot star and a cooler star. Notice that the intensity and the peak frequency increase as the temperature increases.

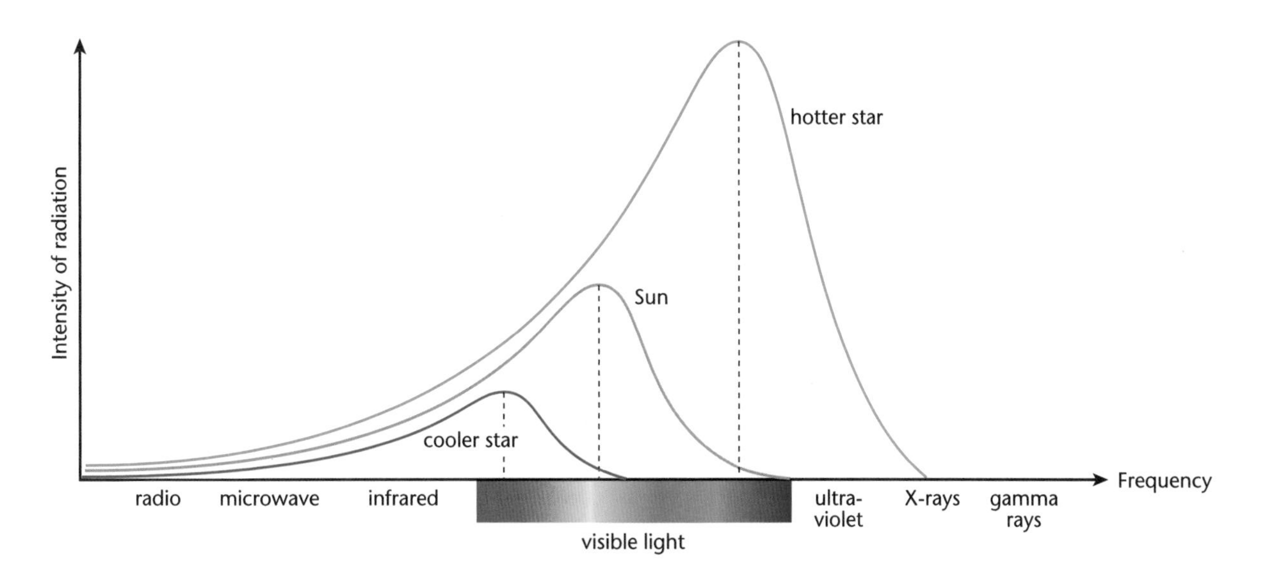

Fig. 12.6 The radiation from different temperature stars (sketch graph).

Cepheid variable stars

OCR A P7.3

Cepheid variable stars can be used to **measure intergalactic distances**.

The star called **Delta Cephei** pulses in brightness. Fig. 12.7 is a graph of how the observed brightness changes over several days. The period of the **variation in brightness** is about 5.3 days.

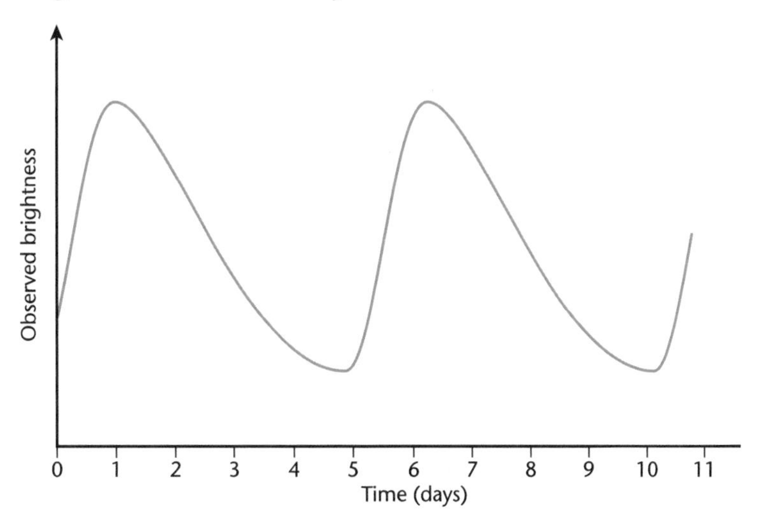

Fig.12.7 The variation in brightness of the star Delta Cephei.

The astronomer Henrietta Leavitt discovered a number of these pulsing stars, called **Cepheid variable stars**, in the galaxies called the Magellanic clouds. The Cepheids she was studying were all in the same distant galaxy – so they were all the same distance from Earth. She realised that the brighter the Cepheid, the longer the period of variation in brightness. She plotted a graph like Fig. 12.8 of **intrinsic brightness** against the **period of variation** of brightness. This gave us a way of working out the intrinsic brightness from the period of variation – and we can measure the period from Earth.

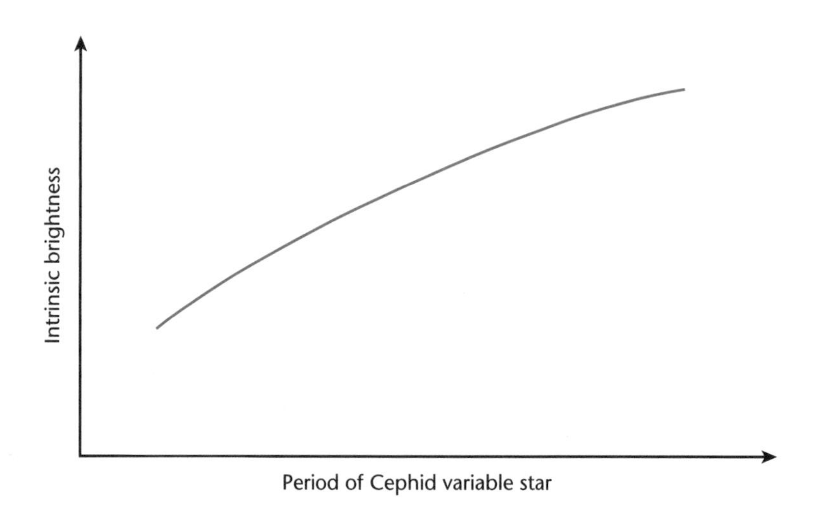

Fig. 12.8 The intrinsic brightness of a Cepheid variable star increases with the period of variation of brightness.

When astronomers discover a new galaxy they can work out how far away it is by:

1. Finding a Cepheid variable star in the galaxy and measuring its observed brightness.
2. Working out its period of variation of brightness by plotting a graph like Fig. 12.7.
3. Using the graph in Fig. 12.8 to work out its intrinsic brightness.
4. Using the intrinsic brightness and average observed brightness to work out the distance to the star (and the galaxy).

Galaxy or Nebula

OCR A P7.3

In remote places where the sky is very dark you can see the **Milky Way** – a brighter strip of the night sky. Observations with telescopes revealed that the Milky Way consists of many stars. Astronomers realised that our Sun was a star in the Milky Way **galaxy**. Telescopes also revealed fuzzy objects in the night sky, and these were named **nebulae** (meaning clouds.) The main issue in the **Curtis-Shapley debate** of 1920 was whether nebulae were objects inside the Milky Way galaxy or separate galaxies outside it (see How Science Works Chapter 5).

Distant galaxies

Edwin Hubble used Cepheid variable stars in the Andromeda Nebula to show that it was a million light years away – which put it much further away than any star in our galaxy, the Milky Way. This proved that it was a separate galaxy – the Andromeda Galaxy, not a cloud of dust (a nebula) in the Milky Way. Cepheid variable stars were very important in determining the size of the Universe, and showing that most nebulae were distant galaxies.

1 megaparsec = 1 million parsecs

Intergalactic distances are typically measured in megaparsecs (Mpc).

The Hubble constant

OCR A P7.3

Hubble's Law was introduced in section 5.6 The Big Bang, and relates the speed at which a galaxy is moving away from us to its distance away.

KEY POINT

Hubble's Law can be written as an equation:

Speed of recession (km/s) = Hubble constant (km/s per Mpc) × distance (Mpc)

Data from Cepheid variable stars in distant galaxies have been used to calculate the **Hubble constant**.

12.4 The life of stars

Finding out what are stars made of

OCR A P7.4

We cannot travel to stars – the nearest after the Sun is about four light years away, at 1000 mph it would take 2.7 million years to get there.

The only information we have about stars comes from the radiation that reaches us from the star. When astronomers realised how far away the stars were they thought they would never find out what they were made of. Then scientists discovered that atoms could be identified from the light they emit, or absorb.

Line spectra

Atomic structure is covered in section 9.1. Photons are covered in section 8.4 The Photon Model

The orbital **electrons** of an atom can only have certain allowed amounts of energy. In their model of the atom scientists imagine a ladder of certain allowed **energy levels**, starting from ground state, or zero level. When the atom absorbs energy an orbital electron moves to a higher energy level. The electron is now said to be in an excited state, and it may emit the energy to return to a lower state or the ground state. The **electron emits** the energy as a **photon** of electromagnetic radiation. The **frequency** of the **photon** depends on the **energy** it transfers .

The electrons in an atom can only exist in certain allowed energy levels. They can only move up/down to a higher/lower allowed level by absorbing/emitting one photon with exactly the right amount of energy – exactly the right frequency.

Fig. 12.9 shows the energy levels of an atom and electrons emitting and absorbing photons. Only a photon with energy that matches the energy difference between levels is absorbed.

Once the energy is enough to ionise the atom, the electron can have any amount of energy – it is no longer part of the atom.

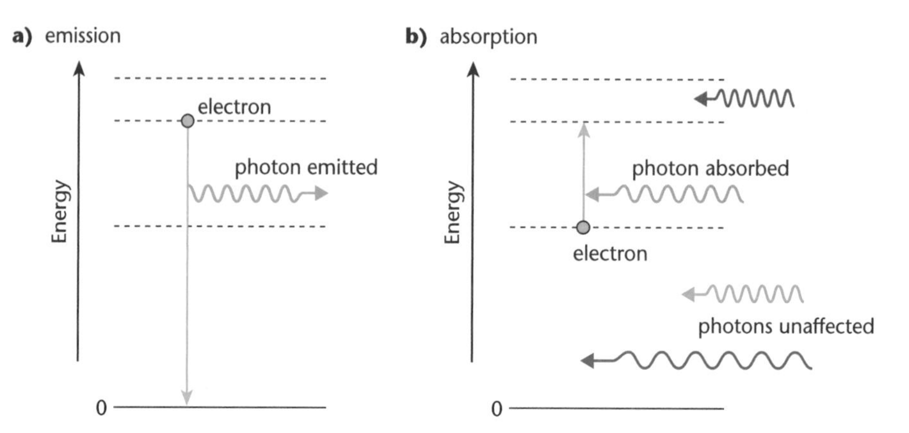

Fig. 12.9 The energy levels of an atom and an electron a) emitting and b) absorbing a photon.

The spectrum of light emitted from a sample of gas or vapour (that has been given energy) will be just the allowed frequencies – a series of **spectral lines** called an **emission spectrum**. In the reverse case, when white light is shone at a sample of gas or vapour the only frequencies absorbed will be the allowed frequencies, and this will result in a series of dark lines. This is called an **absorption spectrum**. These are shown in Fig. 12.10.

The allowed energy levels of an atom depend on the nucleus of the atom, so they are different for each element. This means that the emission or absorption spectrum of light from an element can be used to identify the element. It has been compared to a fingerprint of the element.

On this diagram the lines have been drawn wider to make them clear – on a photograph they would be narrow.

Fig. 12.10 Emission and absorption spectra.

KEY POINT

The spectrum of a star contains some spectral lines and these can be used to identify some of the elements present in the star.

When spectral lines were found in the spectrum of sunlight these were matched with the spectra of known elements. It was clear that the **Sun** contained a lot of **hydrogen**. There were spectral lines that could not be identified and Norman Lockyer suggested this was a new element. He called it **helium** (*Helios* means 'Sun'). When helium was discovered on Earth its emission spectral lines did match the unidentified absorption spectrum lines in the sunlight.

Mystery of the Sun's energy source

It was a mystery to scientists how the Sun had enough fuel to keep burning. They wondered when it would run out of fuel. The discovery of **nuclear fusion** in the early 20th century explained how stars could continue burning for millions of years. You need to know about nuclear fusion, (explained in section 9.3) and how this works in stars (summarised in section 5.4).

The early Universe contained, at first, only **hydrogen** and then **helium** was formed. All the other **elements** were made **inside stars**. They were spread through the Universe as stars ended their lives as **supernova** (an exploding star).

For OCR A you need to think about the properties of gases (temperature and pressure) in section 10.1.

For AQA you need to know that electromagnetic radiation also exerts a pressure, depending on temperature. In large stars the radiation pressure balances the gravitational force and stops contraction.

This section adds more detail to the information in section 5.4.

Star formation

A star forms when the force of gravity pulls a cloud of dust and gas (mostly hydrogen) together to form a protostar. The gas cloud will keep contracting, so the gas is compressed, until the force of gravity is balanced by the forces of the particles colliding. As the pressure increases the temperature increases. The material at the centre starts to glow, and eventually it is hot enough and nuclear fusion starts.

Small masses of dust and gas that contract may be attracted to larger ones as planets, or material further from the cloud centre left over when the star forms may contract to form a planet.

Star endings

Fusion takes place in the **core** of the star. When all the hydrogen in the core has fused to helium the star begins to cool. It stops being a main sequence star. The core **contracts** (due to gravity) because there is less pressure and the outer layers of the star fall inwards, heating up. The hydrogen left in the outer layers starts fusing and the outer layers **expand** and **cool** and the star becomes a **red giant**. The core is still contracting and heats up until **helium** nuclei get close enough to **fuse** into **carbon** nuclei. The more **mass** the star has, the larger the gravitational forces and the hotter it is possible for the core to get. This means the more electrical repulsion can be overcome and the larger the nuclei that can be formed, releasing energy.

A helium nucleus can be added to a carbon nucleus to give **oxygen**. **Nitrogen** is also formed in a later fusion process (and it is very stable so there is a lot of it left in stars).

Only very massive stars become **red supergiants**. They have enough mass to increase the pressure in the core to a higher level (and this increases the temperature). This means other fusion reactions take place, producing all the other **elements** with proton numbers up to the **most stable**, which is **iron** with 26 protons.

A star

Fig. 12.11 shows a cross-section through a star like the Sun and a more massive star.

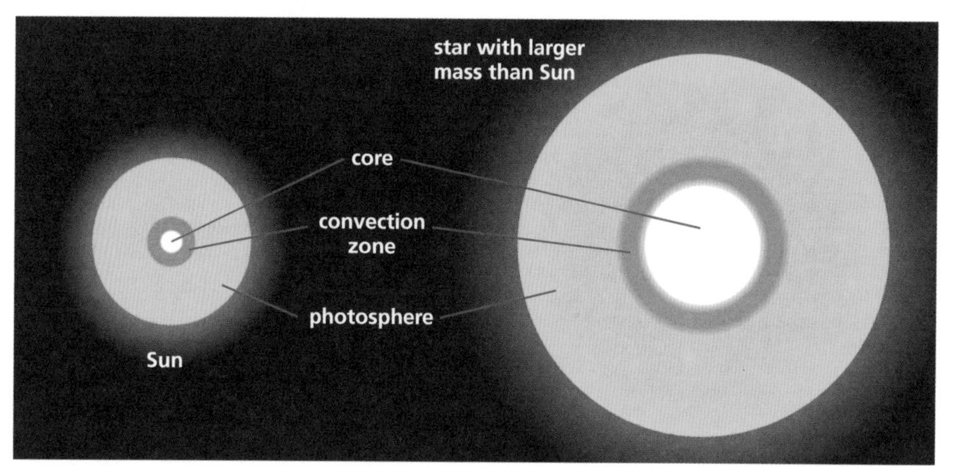

Fig. 12.11 A slice through a star.

A star has:

- a **hot core** where fusion takes place
- a **convective zone** where energy is transported to the surface by convection currents
- a **photosphere** where electromagnetic radiation is emitted into space.

12.5 Lenses, mirrors and telescopes

Converging lenses

Lenses make use of refraction (see section 3.5) to change the direction of light rays. A **converging lens** makes a **parallel** beam of light rays converge to a point, called the **focal point** or **principal focus**. All converging lenses are fatter in the middle than at the edge.

For OCR B you need to know that the converging lenses shown here are also called convex lenses.

KEY POINT **The focal length, *f*, of a lens is the distance from the centre of the lens to the point where a parallel beam of light is focussed to a point. (The point is called the focus, focal point or principal focus.)**

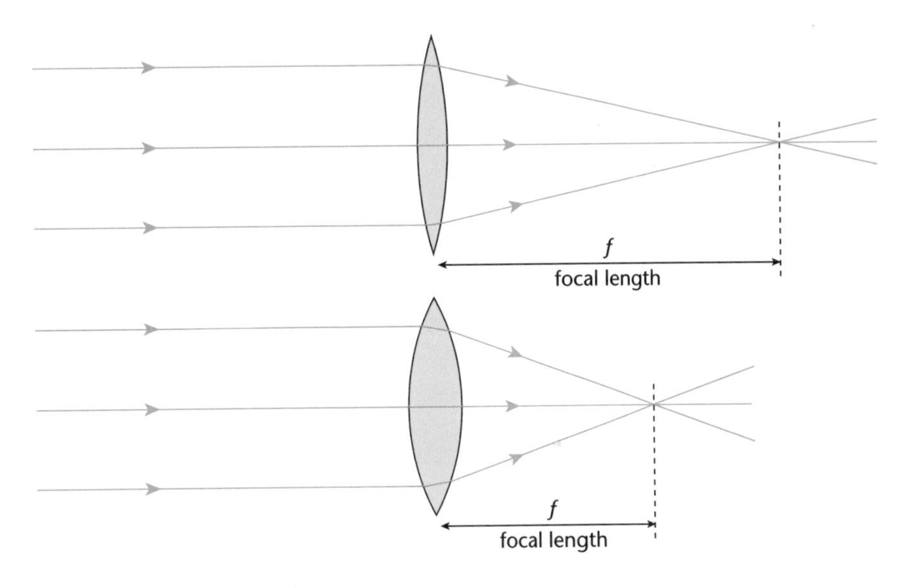

Fig 12.12 Converging lenses with different focal lengths.

For two lenses made of the same material, the one with the most curved surfaces will have the shortest focal length.

Diverging rays

Light from a source spreads out, or diverges. A converging lens can be used to reduce the divergence, as shown in Fig. 12.13.

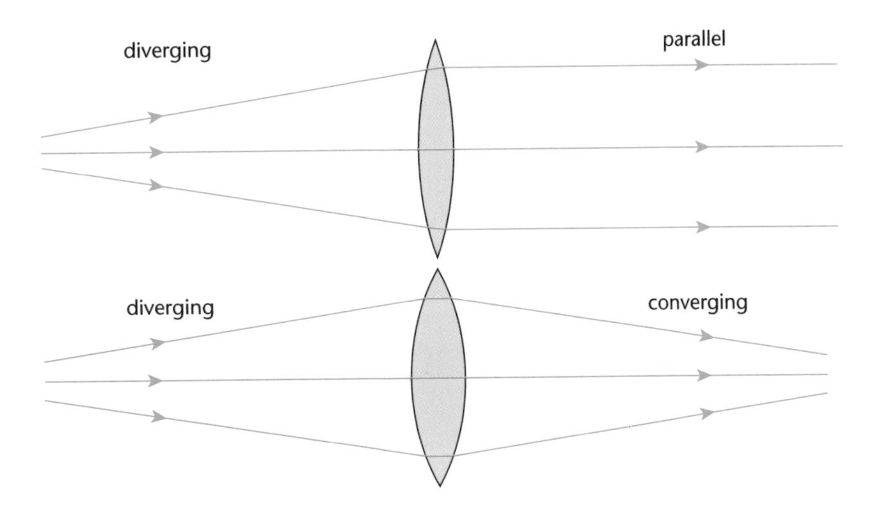

Fig. 12.13 Using a converging lens to make a parallel or converging beam.

Images

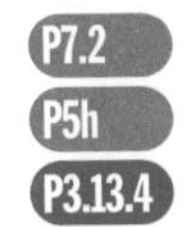

Describing the image

A lens is used to produce **images** in telescopes, cameras, projectors, magnifying glasses and other devices. The image is a copy of the **object**. Table 12.1 lists the words used to explain what an image is like.

To decide if the image is real it may help to think, 'Can I touch it?'

Image is:	This means:
real	It can be displayed on a screen because the light passes through it
virtual	It can only be seen by looking through a lens – the light doesn't pass through it
upright	The same way up as the object
inverted	Upside down compared to the object
magnified	Bigger than the object

The image you see of yourself in a mirror is a virtual image.

A converging lens can form real, inverted images or virtual, upright images depending how far it is from the object.

Magnification

The magnification of the image depends on the distance between the object and the lens, and on the focal length of the lens.

KEY POINT

The magnification produced by a lens is:

$$\frac{\text{The image height}}{\text{The object height}}$$

Magnification has no units.

Example

If the image is twice the length of the object then the magnification is:

$$\frac{2}{1} = 2$$

If the image is 0.5 cm and the object is 2 cm. The magnification is:

$$\frac{0.5\,\text{cm}}{2\,\text{cm}} = 0.25$$

Drawing ray diagrams

You can find out what in image is like – and where it appears – by drawing a ray diagram. To find a real image, the method makes use of two key points:

1 Light passes straight through the centre of a lens.

2 Light parallel to the axis is converged so it passes through the focus, (or light passing through the focus is converged parallel to the axis).

These are the steps involved:

- Decide a suitable scale.
- Draw a vertical line to represent the lens and a horizontal line to represent the axis through the centre of the lens.
- Draw an upright arrow to represent the object – this must be to scale.
- Mark the focus – also to scale
- Draw a line from the top of the arrow straight through the centre of the lens
- Draw a line from the top of the arrow parallel to the axis until it meets the lens. Continue the line on the other side of the lens so it passes through the focus.
- Where the two lines cross is the top of the image, the bottom will be on the axis. Draw the image arrow

Ray diagrams are best drawn on graph paper with a sharp pencil and a transparent ruler (so you can see the whole diagram).

For OCR A, you will use this method only to draw diagrams for the parallel light from stars or galaxies.

Fig. 12.14 shows some ray diagrams drawn using this method.

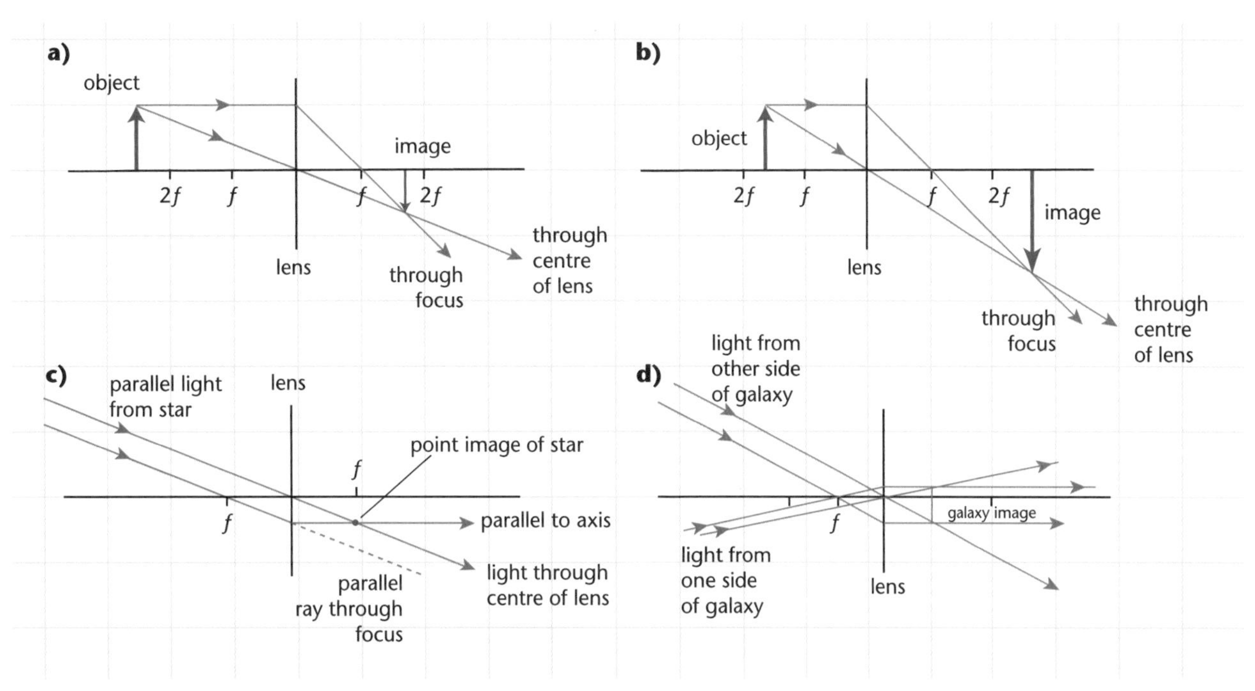

Be able to draw:
For AQA: a) and b)
For OCR B: a) and b)
For OCR A: c)

Fig. 12.14 ray diagrams.
a) For a camera, the object is a long way from the lens and the image is smaller
b) For a projector, the object (slide or film) is closer to the lens and the image is magnified
c) Star-light comes from so far away it is parallel
d) Parallel light from both sides of a galaxy gives an inverted image of the galaxy.

The magnifying glass

The magnifying glass

To use a converging lens as a magnifying glass (as shown in Fig. 12.15) the lens must be held closer to the object than the focal length. The image is virtual, upright, and magnified. You must look through the lens to see it. The image appears to be further away than the object.

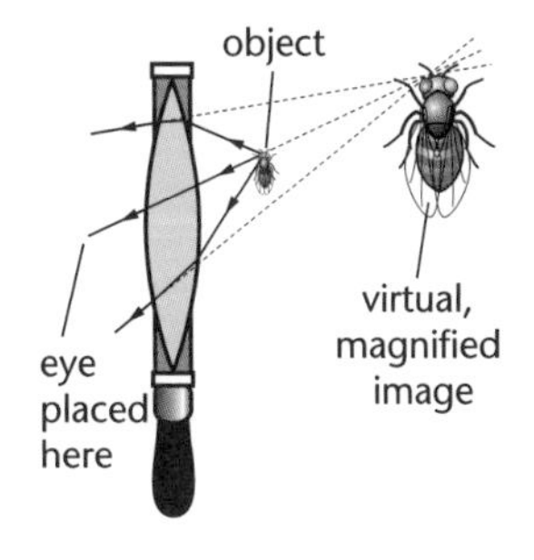

Fig. 12.15 A converging lens as a magnifying glass.

The camera

OCR B P5h
AQA P3.13.4

The converging lens in the camera produces a real inverted image that is smaller than the object. The lens must be more than twice the focal length from the object so that it produces a small image. To focus the image the camera lens is moved:

- away from the film to focus on a close object
- towards the film to focus on a distant object.

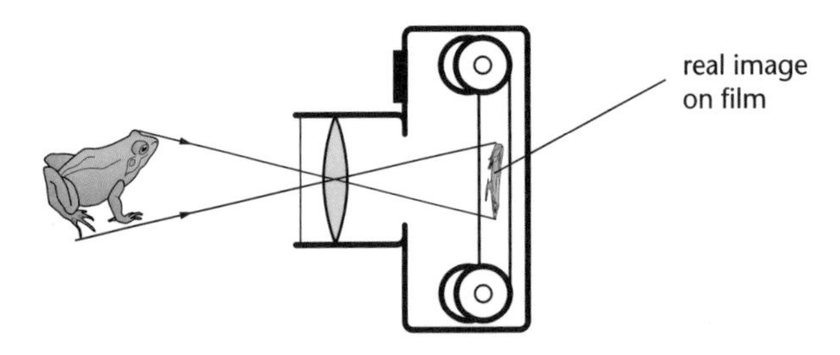

Fig. 12.16 A converging lens in a camera.

The projector

OCR B P5h

Projector

The converging lens in the projector produces a real inverted image that is magnified. The lens must be between one and two times the focal length from the slide or film so that it produces a magnified image. To focus the image the projector lens is moved:

- away from the slide to focus on a closer screen and give a smaller image
- towards the slide to focus on a distant screen and give a more magnified image.

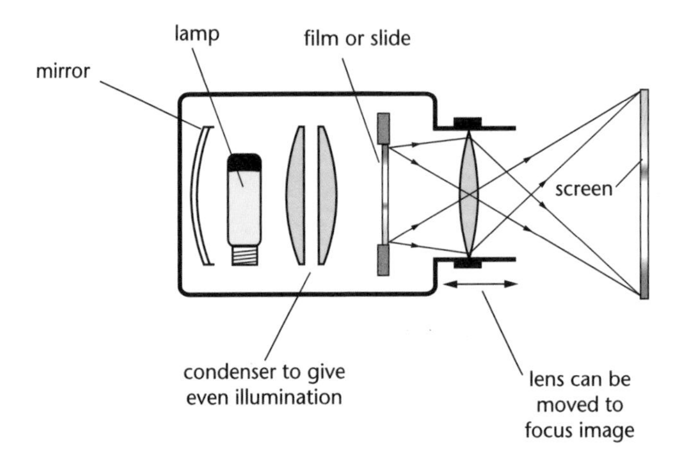

Fig. 12.17 A converging lens as a projector.

More about virtual images

AQA P3.13.4

Virtual image from a converging lens

To produce a virtual image using a converging lens, the object must be between the principal focus and the lens. Table 12.2 summarises the image position for different object positions for a converging lens.

Object position (focal length of lens = f)	Image position	Properties of image and uses
More than 2*f* from lens	Between *f* and 2*f* from lens	Real, inverted, smaller than object Cameras
2*f* from lens	2*f* from lens	Real, inverted, same size as object
Between *f* and 2*f* from lens	More than 2*f* from lens	Real, inverted, magnified Projectors
f from lens	Parallel beam – image 'at infinity'	When object is a light source lens gives a parallel beam
Less than *f* from lens	Same side of lens as object, further away from lens than object	Inverted, upright, magnified Magnifying glass

Fig. 12.18 shows how to draw a ray diagram when the lens is used as a magnifying glass. To find the position of the image dotted lines are used to show that the light appears to have come from the direction of the dotted line.

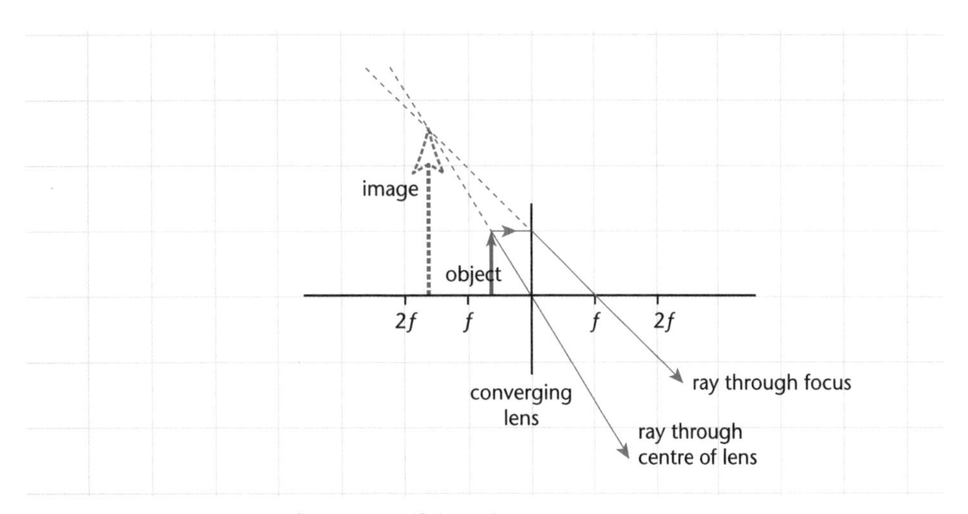

Fig. 12.18 Ray diagram for a magnifying glass.

Diverging lenses

Fig. 12.19 shows a diverging lens. The lens makes a parallel beam diverge so that it appears to have come from the focus. Diverging lenses produce virtual images that are upright and smaller than the object. Fig. 12.20 shows how to draw a ray diagram to find the position of the image.

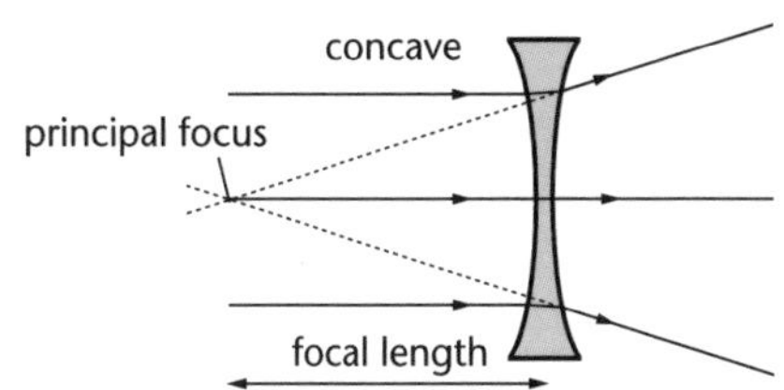

Fig. 12.19 A diverging lens.

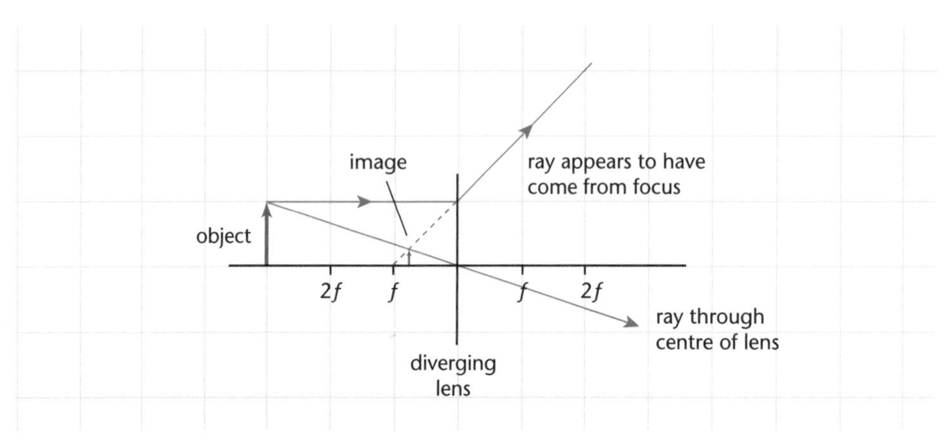

Fig. 12.20 Ray diagram for a diverging lens.

Refracting telescopes

Telescopes made with lenses are called refracting telescopes. You can make one with two converging lenses, the one nearest your eye is called the **eyepiece lens** and the one nearest the object is called the **objective lens**. The lenses are arranged so that the objective lens produces a real image and the eyepiece lens is used as a magnifying glass to magnify this image.

The power of a lens

The shorter the focal length, the more powerful the lens.

KEY POINT

The power of a lens (in diotres, D) = $\frac{1}{\text{focal length (in metres, m)}}$

Example

If a lens has a focal length of 0.8 m it has a power of 1 ÷ 0.8 = 1.25 D

A lens with a focal length of 10 cm has a power of 1 ÷ 0.1 = 10 D

Converging lenses have positive powers (for example +2 D) diverging lenses have negative powers (for example –0.5 D)

Choosing the lenses

To work as a telescope, the eyepiece lens must be the more powerful lens – the fatter, or more curved lens. The objective lens is the thinner lens.

KEY POINT

magnification = $\frac{\text{focal length of objective lens}}{\text{focal length of eyepiece lens}}$

Example

A telescope has an eyepiece lens with power 2 D and an objective lens with power 0.25 D.

Focal lengths are $f_e = 1 \div 2\,\text{D} = 0.5\,\text{m}$ and $f_o = 1 \div 0.25 = 4\,\text{m}$

Magnification = $f_o \div f_e = 4\,\text{m} \div 0.5\,\text{m} = 8$

(or notice that magnification = $\dfrac{\text{power of eyepiece lens}}{\text{power of objective lens}} = 2\text{D} \div 0.25\,\text{D} = 8$)

Stars are so far away they still look like points through a telescope, but groups of stars will be spread out

Looking through this telescope makes the Moon appear 8 times (or 8×) bigger than without the telescope, but it is not true to say the size of the image is 8 times bigger than the real size of the Moon. A magnification of 8 × makes the Moon appear 8 times nearer, or we can say the telescope has an **angular magnification** of 8 × . This means that the angle between rays from the top and bottom of the Moon is 8 times bigger, as shown in Fig. 12.21.

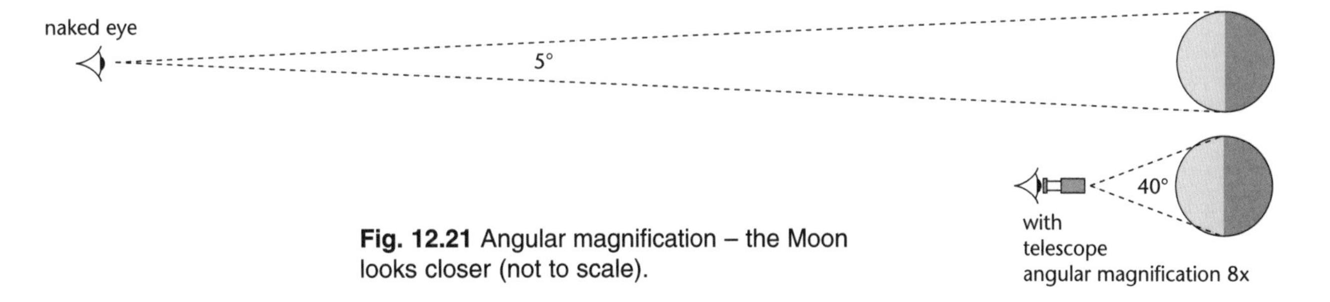

Fig. 12.21 Angular magnification – the Moon looks closer (not to scale).

Aperture and brightness

The pupil of your eye is very small. The objective lens of the telescope gathers all the light rays that enter it, and focuses them to a point. This will make the star brighter. You can see stars that are too dim to see with the naked eye. The diameter of the objective lens is called the **aperture**. To collect radiation from weak or distant sources you need a telescope with a large aperture.

Another reason for using a large aperture is to reduce **diffraction effects**. This is especially important for longer wavelengths, see section 8.3.

Mirrors

AQA P3.13.4

Plane mirrors

The reflection at a plane (flat) mirror is covered in section 3.5. Fig. 12.22 shows how to draw a ray diagram to show where the image appears.

- Draw rays from the object to the mirror surface.
- Draw a normal at the mirror surface.
- Draw a reflected ray so that the angle of incidence i = the angle of reflection r .
- Extend these rays back to the point where they meet – this is the location of the image.

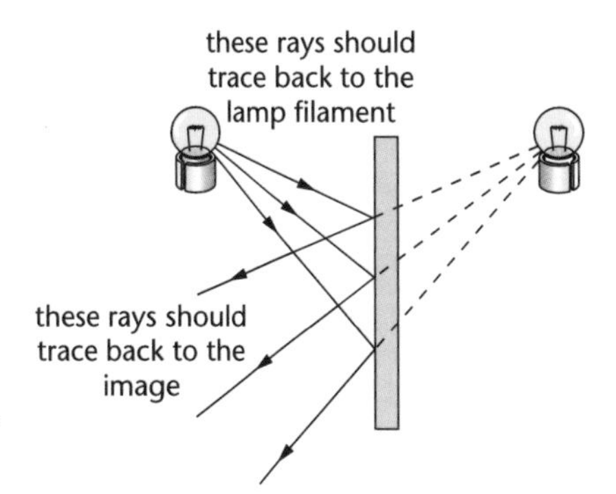

Fig. 12.22 The image in a plane mirror.

The image in a plane mirror is:
- virtual (behind the mirror)
- the same size as the object
- the same distance behind the mirror as the object is in front
- upright
- laterally inverted – the left-hand side in the mirror is the right-hand side of the object.

Curved mirrors

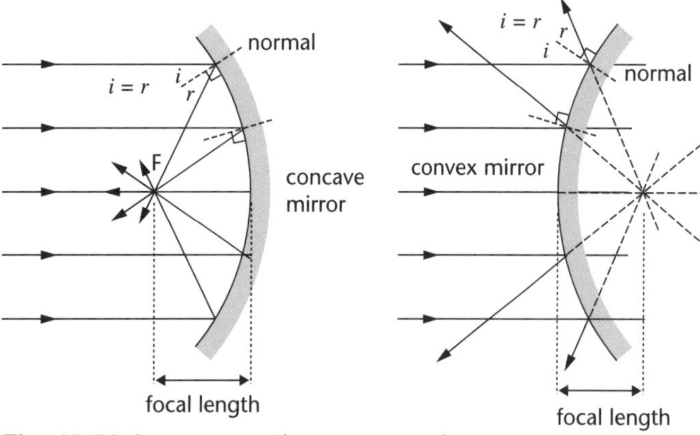

Fig. 12.23 A convex and a concave mirror.

These still obey the **law of reflection**, but, because the **normal** has a different direction at different points on the mirror surface, they can produce magnified and smaller images. The **focus** is the point where:
- a **concave mirror** focuses rays of parallel light to a point
- the parallel rays diverged from **convex mirror** appear to have come from.

The **focal length** is the distance from the focus to the mirror surface.

The image in a concave mirror

The image is different depending on how close the object is to the mirror. To draw a ray diagram you need to know:
- a ray parallel to the axis is reflected through the **focus**
- a ray through the focus is reflected parallel to the mirror
- a ray that passes through the **centre of focus** (C) is reflected straight back (this is because it is along the normal).

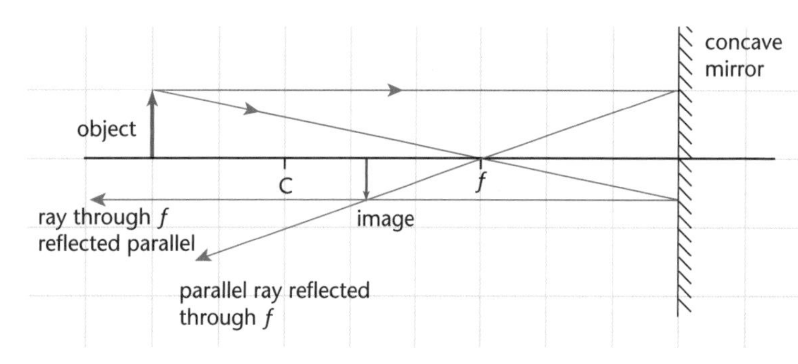

Fig. 12.24 shows how to draw a ray diagram for one position of the image. As with a lens the rays will converge at different positions and give different types of image depending on the position of the object. This is shown in the table below.

Fig. 12.24 The image in a concave mirror.

Object position C= centre of curvature F = focus	Image position	Type of image
Further away than C	Between C and F	Real, inverted, smaller
At C	At C	Real, inverted, same size
Between C and F	Further away than C	Real, inverted, magnified
At F	A parallel beam, at infinity	A light source placed at F will give a parallel beam
Between F and mirror	Behind the mirror	Virtual, upright, magnified

The image in a convex mirror

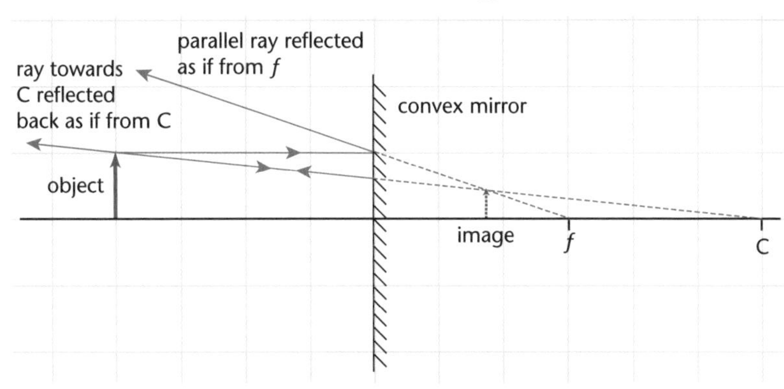

Fig. 12.25 The image in a convex mirror.

This is always virtual (behind the mirror,) upright and smaller than the object. To draw a ray diagram you need to know:

- a ray parallel to the axis is reflected as if it has come from the **focus**
- a ray that appears to go to the **centre of curvature** (C) is reflected straight back (this is because it is along the normal).

Reflecting telescopes

OCR A P7.2

> **KEY POINT** **Most astronomical telescopes have concave mirrors, not converging lenses, as their objectives.**

There are lots of different designs, but all have a very large mirror to focus the light gathered. The eyepiece is then positioned to look at the focused light.

The mirror is a **concave** shape, but it is not part of a sphere, it is **parabolic**. This is so that all the rays are focused at a single point, as shown in Fig. 12.26.

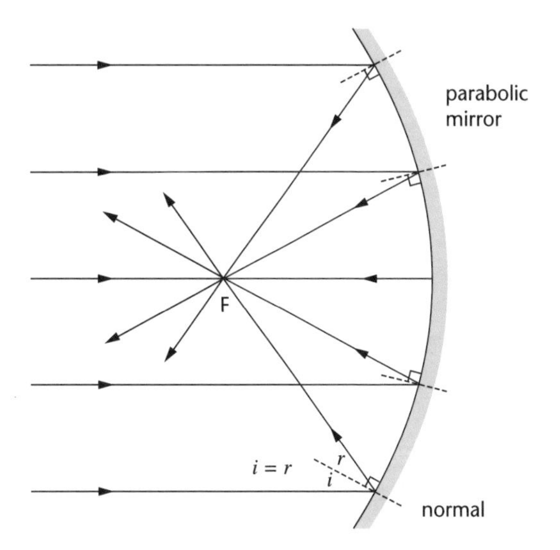

Fig. 12.26 A concave reflector for a telescope.

The advantage of using a mirror is that you can make very large mirrors and support them from behind, but you cannot make such large lenses because:

- the lens would be so heavy it would distort under its own weight
- it would be difficult to make the glass even so that it had the same refractive properties at every point.

HOW SCIENCE WORKS

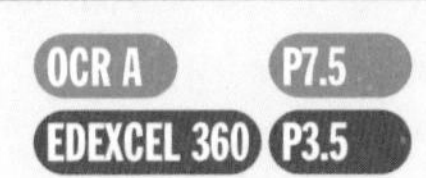

How astronomers work together

There are many observatories all over the world.

Jodrell Bank Observatory in the North West of the UK has a large radio telescope called the Lovell Radio telescope, and is also the home of the MERLIN National Facility. MERLIN is a group of telescopes in the UK, including the Lovell Telescope, which can all be operated together. The Lovell telescope, shown in Fig. 12.27, is 76 m in diameter and can be turned to point at the area of the sky to be investigated. The detectors at the focus of the large parabolic dish pick up radio waves at five different wavelengths ranging from 1.4 cm to 20 cm. The aperture needs to be large to pick up more radiation and to reduce diffraction effects.

Fig. 12.27 The Lovell radio telescope.

By operating a group of telescopes together the aperture is equal to the area of all of them added together. The Lovell telescope is also regularly linked to telescopes in Europe and around the world to make observations. They are used to look at both nearby stars in the Milky Way and the most distant galaxies.

Radio waves pass through the atmosphere, so radio telescopes are still sited in the UK, but visible light wavelengths are affected by absorption and refraction in the atmosphere, and also by light pollution from cities, so astronomers have moved their world-class telescopes to better sites. They need remote sites, so there is no artificial light, and mountain-tops to get above as much of the atmosphere as possible. It is important to consider the weather conditions – to maximise the nights the telescope can be used.

A number of international observatories have been built, 2400 m above sea level, on Roque de los Muchachos in the mountains of the Canary Island, La Palma. This site was chosen because the clouds usually form between 1000 m and 2000 m above sea level, below the telescopes. Telescopes here operate at visible wavelengths and some infrared wavelengths.

The Hubble telescope is the most famous satellite telescope, but there are others. If you want to observe X-ray wavelengths, or some infrared wavelengths you need an orbiting telescope outside the atmosphere because these wavelengths are absorbed by the atmosphere.

HOW SCIENCE WORKS

The table below summarises the advantages and disadvantages.

Orbiting telscopes, e.g. Hubble Telescope and the X-ray Multi-Mirror Newton telescope	
Advantages	**Disadvantages**
Avoids absorption and refraction effects of the atmsosphere	High cost of setting up (launch), maintaining and repairing
Can use parts of the electromagnetic spectrum that the atmosphere absorbs	Can be affected by problems with the space program (e.g. launch cancelled due to rocket design problems)

The European Southern Observatory in the mountains of Chile has three observatories set up by 11 European countries. International collaboration between astronomers saves money and means they can share ideas and expertise. The site is very remote, and instead of travelling there to make observations, the astronomers make use of computers. The telescopes are all under remote control and instructions are sent from, and data returned to, the control centre in Germany. Astronomers download their data from the internet. Computers can position the telescopes very accurately and handle the vast amount of data that is collected.

This doesn't mean that there are no staff at the observatories. Some staff are needed to keep things running smoothly, and for maintenance. There are residential buidings with some rooms for visiting astronomers and the observatories provide a number of jobs for local people. Important factors (other than astronomical ones) to consider when planning, building, operating, and closing down an observatory are:

- cost
- the impact on the environment, and on people living near the observatory
- working conditions for employees.

1. Which of these statements are true?
   - **(a)** Radio telescopes are not put in orbit because it would be too expensive.
   - **(b)** X-ray telescopes are always put in orbit because X-rays are dangerous to living things.
   - **(c)** Four identical telescopes can be used to collect visible light and the combined data is the same as having one telescope with four times the aperture.
   - **(d)** Wales would be a good place to site an international observatory because it is remote. **[4]**
2. Give two examples of the locations of major observatories. **[2]**
3. Give an advantage and a disadvantage of putting a telescope for visible wavelengths in orbit. **[2]**
4. A group of astronomers are choosing a site for an observatory. Suggest:
   - **(i)** an important astronomical factor to take into account
   - **(ii)** an important non-astronomical factor to take into account. **[2]**

Exam practice questions

1. A magnifying glass is used to look at a crystal that is 3 mm wide. The magnification is ×4. How wide is the image?

A 3 mm
B 4 mm
C 7 mm
D 12 mm **[1]**

2. Inside stars energy is released by:

A helium nuclei fusing to give hydrogen nuclei
B hydrogen nuclei fusing to give helium nuclei
C hydrogen nuclei fusing to give all the elements
D helium and hydrogen nuclei fusing with fast neutrons **[1]**

3. Jack is making a telescope using two lenses.

Lens A has a power of +1.25 D and a diameter of 4 cm
Lens B has a power of + 2.5 D and a diameter of 5 cm

Write **T** for the **True** statements and **F** for the **False** statements:

(a) Lens A should be used as the eyepiece lens and lens B as the objective lens.
(b) The focal length of lens B is 40 cm.
(c) The magnification of the telescope is × 2.
(d) With the naked eye, the Moon has an angular size of about 0.5°. With this telescopethe Moon has an angular size of about 2°
(e) Through the telescope the Moon appears upside down (inverted).
(f) Increasing the aperture of the telescope will make the image of the Moon brighter. **[3]**

4. A ray diagram can be used to show how a lens produces an image of an object.

Choose from this list the **three** things you have to know to be able to draw a ray diagram to find out the **height of the image** and **how far it is from the lens**:

A focal length of the lens
B thickness of the lens
C height of the object
D distance of the object from the lens
E which way up the image is **[3]**

Exam practice questions

5. Use words in the list to complete the sentences describing how to work out the distance to a galaxy. (Use words once, more than once or not at all.)

brightness **Cepheid** **dwarf** **giant** **intrinsic**
observed **period** **red** **variable** **white**

Observe the galaxy and locate a ___1___ ___2___ star. Measure the ___3___ ___4___ of the star and its ___5___. Use a graph plotted using data from lots of these stars with ___6___ ___7___ on the vertical axis and ___8___ on the horizontal axis, to find the ___9___ ___10___ of the star. Work out the distance to the galaxy by comparing the ___11___ ___12___ to the ___13___ ___14___. **[7]**

6. Astronomers have collected some absorption spectra data from a star.

The tables list:

- the wavelengths of the absorption spectral lines in the spectrum of the star
- the emission spectral lines of some elements.

Wavelength of absorption spectral lines (nm)
Star
486
502
517
518
588
589
590
656
668

Wavelength of emission spectral lines (nm)			
helium	**hydrogen**	**iron**	**magnesium**
502	486	467	517
588	656	527	518
668			

Tick (✔) the elements present in the star:

(a) helium?
(b) hydrogen?
(c) iron?
(d) magnesium? **[4]**

7. **(a)** A convex lens is used to produce an image of a light bulb on a screen. The lens is positioned to make the image the same size as the light bulb.

(i) How would you move the lens to make the image of the light bulb larger?

(ii) Give an example of something that uses a convex lens in this way to make a large image.

(b) A digital camera lens is used to make a small image on the light sensitive detectors. The image is out of focus, the photographer adjusts the focus to get a sharp image. What happens inside the camera to focus the light on the light sensitive detectors? **[3]**

Chapter 13

Applications – electronics and medicine

The following topics are covered in this chapter:

- *Using electrical resistance*
- *Diodes and capacitors*
- *Medical physics*
- *Uses of electromagnetism*
- *Digital electronics*

13.1 Using electrical resistance

Circuit basics

In addition to the circuit symbols covered in section 7.2 Electric circuits, you need to know the symbols in this table.

Component	Symbol	Component	Symbol
light emitting diode (LED)		capacitor	
electrolytic capacitor	+	relay	NO COM NC
NOT gate	NOT	AND gate	AND
OR gate	OR	NOR gate	NOR
NAND gate	NAND		

More about resistance

OCR B P6a

This builds on section 7.2 Resistors and resistance.

KEY POINT

The resistance, R, of a component is:

$$R = \frac{V}{I}$$

Where V is the voltage (or potential difference) in volts and I is the current in amperes. Resistance is measured in ohms.

R = gradient of a graph of V against I

Fig. 13.1 shows a graph of voltage V on the vertical axis against I on the horizontal axis for a resistor which has a fixed resistance.

The gradient of the graph $= \frac{V}{I} = 100$

So the resistance of the resistor = 100 ohms

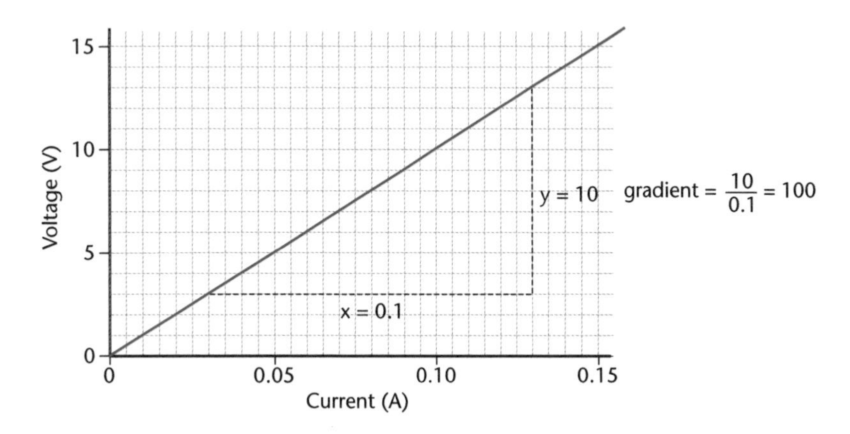

Fig. 13.1 A graph of voltage against current for a 100 ohm resistor.

If you alter the voltage and measure the current, voltage is the independent variable and current is the dependent variable. You would usually plot a graph of current against voltage, as shown in section 7.2. Graphs of **voltage** against **current** have the advantage that the **gradient** is equal to the **resistance**.

Voltage against current graphs can be plotted for components where the resistance varies and the gradient of these graphs shows how the resistance varies. For example, the wire in a **filament lamp** is designed to get hot enough to glow. Its resistance increases as the current increases because it gets very hot. The graph of **voltage against current** looks like Fig. 13.2. If the graph for a component is not a straight line, it is because the resistance is changing and the component is sometimes called a **non-ohmic device**.

It is more usual to plot current against voltage as shown in section 7.2.

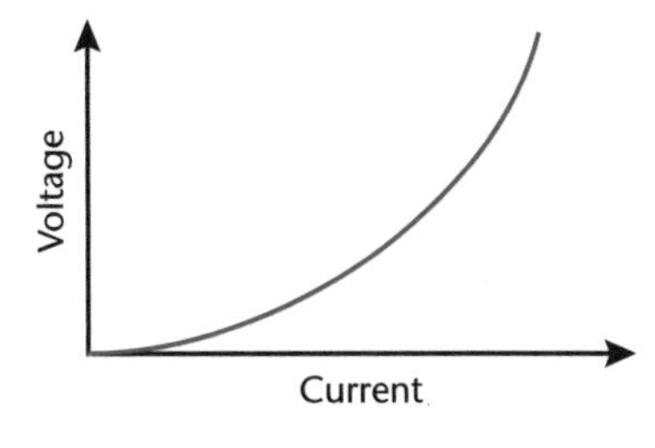

Fig. 13.2 A graph of voltage against current for a filament lamp.

Potential dividers

P6b

A **potential divider** circuit made using two resistors can be used to provide an **output p.d.** with the value that is wanted from a higher **input p.d.** Using the fact that the current in a series circuit is the same everywhere, and that current = voltage ÷ resistance, in the circuit shown in Fig. 13.3:

$$\text{current} = \frac{V_{in}}{(R_1 + R_2)} = \frac{V_1}{R_1} = \frac{V_2}{R_2}$$

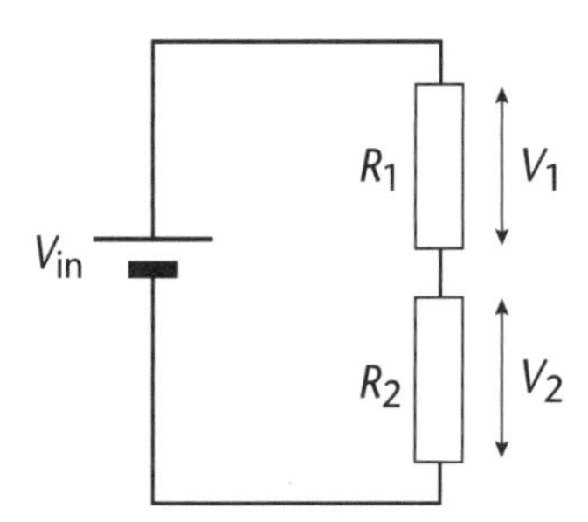

Fig. 13.3 A series circuit, the current is the same through R_1 and R_2.

Fig. 13.4 shows a **potential divider** circuit made using two fixed resistors, R_1 and R_2. In this potential divider circuit:

- the input p.d. is the sum of the p.d.s across the resistors
- the p.d. across the resistors is divided in the ratio of their resistances, so the resistor with the greatest resistance has the greatest p.d. across it.

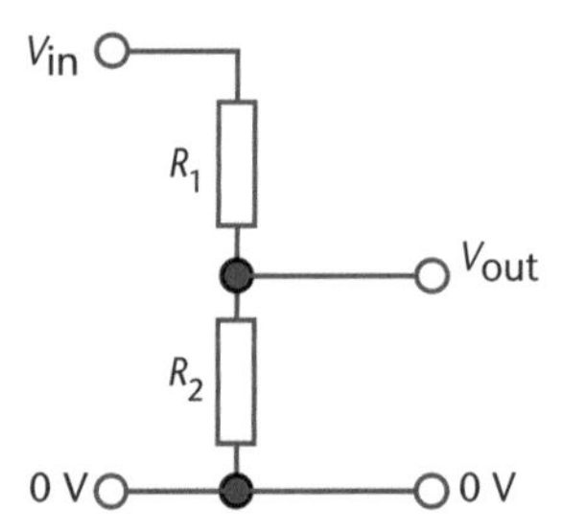

Fig. 13.4 A potential divider circuit.

Make sure you can use this equation to work out the output p.d.

For the potential divider circuit shown in Fig. 13.4:

$$V_{out} = V_{in} \times \frac{R_1}{(R_1 + R_2)}$$

Example

When the input p.d. is 5V, R_1 = 300 ohms and R_2 = 200 ohms.

$$\text{The output p.d.} = 5 \times \frac{200}{(300 + 200)}\text{V} = 2\text{V}$$

A potential divider with variable output

For more about LDRs and thermistors see section 7.2.

Fig. 13.5 shows a circuit with a **variable resistor**, so that it can be used to provide a variable output p.d. from a fixed input p.d. of 5 V.

- When the resistance of the variable resistor is zero, the output voltage equals the input voltage.
- As the resistance of the variable resistor is increased the output voltage is reduced.
- When the resistance of the variable resistor is much larger than R_2 (for example one hundred, or one thousand times bigger) the output voltage is so small that for practical purposes it is zero.

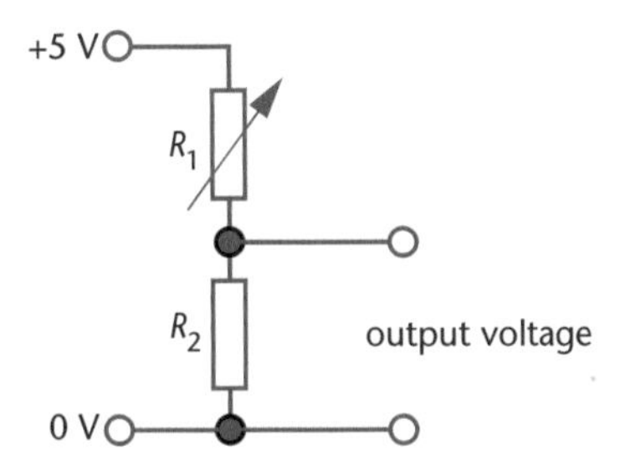

Fig. 13.5 A variable resistor in a potential divider can be used to change the output p.d.

Using thermistors and LDRs

Fig. 13.6a shows a potential divider with a **thermistor**. As the **temperature** increases the resistance of the thermistor falls, so the output p.d. in the circuit falls. This circuit could be used to switch a heater on and off. When the temperature is high enough, the output p.d. drops low enough to switch off the heater. The **threshold** p.d. is the p.d. when the heater switches off. If R_1 is replaced with a **variable resistor** it can be used to adjust the threshold temperature at which the heater switches off.

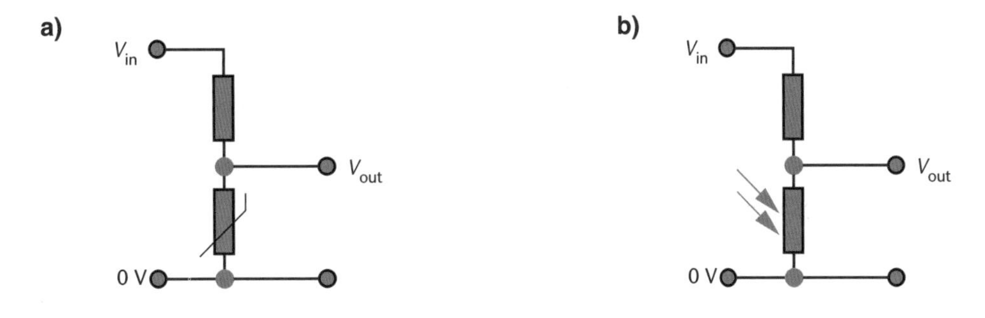

Fig. 13.6 a) A temperature dependent potential divider.
b) A light dependent potential divider.

Fig. 13.6b Shows a similar circuit with a **light-dependent resistor** (**LDR**). When the **light intensity** falls, the resistance increases and the output p.d. increases. This circuit could be used to switch a light on when it gets dark. The **threshold** at which the light switches on could be adjusted by replacing R_1 with a **variable resistor**.

13.2 Electromagnetism

The a.c. generator

OCR B P6d
AQA P3.13.8

This section builds on 7.3 More about electromagnetic induction and section 2.4 How generators work.

When a wire (or other conductor) moves in a **magnetic field** a voltage is **induced** across it. The size of the voltage depends on the rate at which the magnetic field changes.

Large generators in power stations usually rotate an electromagnet to induce voltage in stationary coils, but in the design shown in Fig. 13.7 it is the coil that rotates between stationary magnet poles. (This is the design used, for example, to supply the electricity in cars.) This means that some method of connecting the coil to a circuit is needed that will not get twisted as the coil turns. **Slip-rings** are used which make close electrical contact with **brushes**. The brushes are named because they brush against the slip-rings. They are made of a hard material or they would quickly wear down.

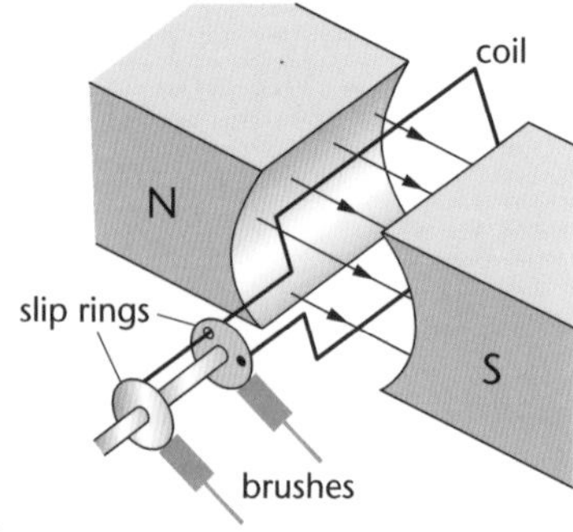

Fig. 13.7 An a.c. generator.

The size of the induced voltage can be increased by:

- increasing the number of turns on the coil
- increasing the area of the coil
- moving the magnet faster
- increasing the strength of the magnetic field.

When the coil is horizontal, a small movement cuts a lot of magnetic field lines and the induced voltage reaches a peak before reducing again. When the coil is vertical, for an instant it moves parallel to the magnetic field and cuts no magnetic field lines, so the induced voltage is zero. It then changes direction –if it was moving up it starts moving down, and vice versa. This is when the induced voltage changes direction and current flows the opposite way around the coil and the external circuit.

Fig. 13.8 is a graph of the alternating voltage, showing how it changes when the rotation is speeded up.

You will find details of transformers in section 7.3 the mains electricity supply.

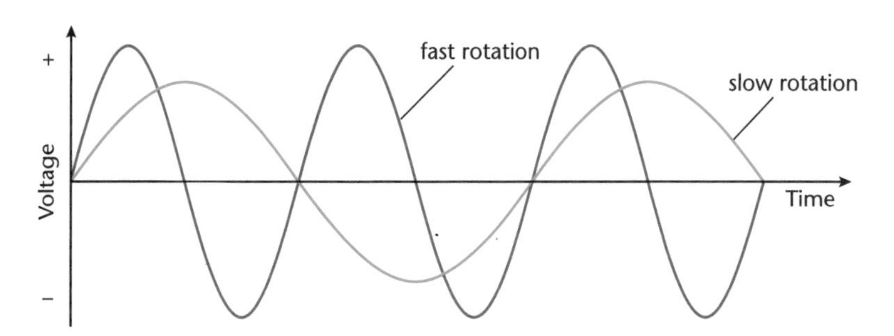

Fig. 13.8 Alternating voltage for different coil rotation speeds.

Electric current and magnetic fields

For OCR B, you may be asked to describe the shape of the fields.

When an electric current flows in a conductor. a **magnetic field** is set up around it. Fig. 13.9 shows the shape of these fields for a straight wire, a coil, and a long coil called a **solenoid**. The magnetic field is:

- increased when the electric current is increased
- reversed when the electric current is reversed. (The direction of a magnetic field is from the north pole of a magnet to a south pole.)

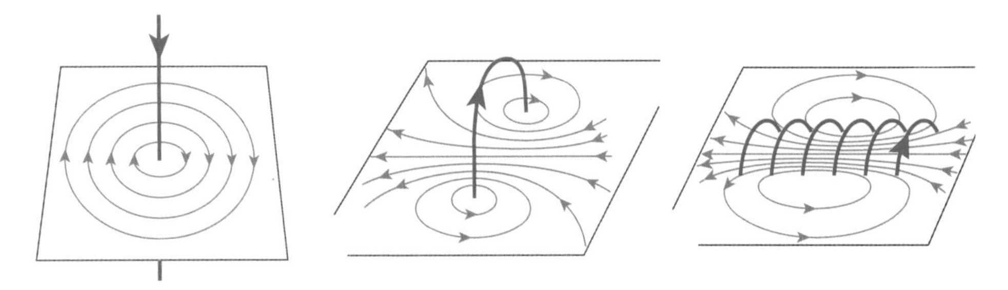

Fig. 13.9 The magnetic field around a current carrying wire, coil and solenoid.

The coil (or wire), is now an electromagnet and, in the right position, it will attract or repel another magnet. This force is made use of in many electromagnetic devices, including the **electric motor**, and moving the cone of a loudspeaker. The effect is called the **motor effect**. The coil (or wire), must be positioned so that the **current** is flowing in a direction **perpendicular** to the **magnetic field** direction. (When the field and the current are parallel there is no force.)

To work out the direction of the force, and movement, you can use **Fleming's Left-Hand Rule**. (If you know the direction of movement, use it to find the direction of the current or the field.)

- The **direction** of the force (and movement) can be reversed by reversing the direction of the **magnetic field** OR the **electric current**.
- The **size** of the force (and movement) can be increased by increasing the size of the **magnetic field** AND/OR the **electric current**.

It is important to use the left hand – the right hand gives the wrong direction.

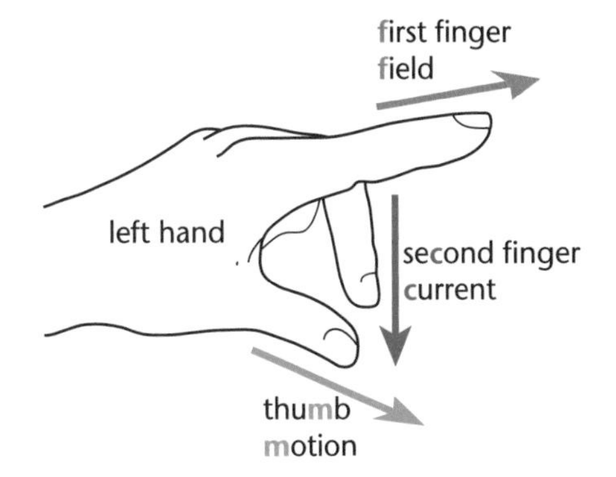

1. point your first finger in the direction of the magnetic field.
2. point your second finger in the direction of the current.
3. your thumb points in the direction of motion.

Fig. 13.10 Using Fleming's Left-Hand Rule to find the direction of the motion.

The electric motor

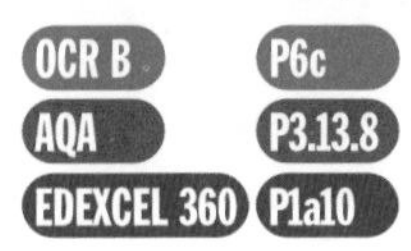

Fig. 13. 11 shows a simple d.c. electric motor. A rectangular coil is free to turn on an axle. It is positioned in a magnetic field, between the poles of two magnets, so that, when the coil is horizontal and a current flows in it, there is a force on each side of the coil. Fig. 13.12 shows that the force is up on one side of the coil and down on the other. These two forces have a turning effect on the coil. When it reaches the vertical position, the direction of the force must be reversed to make the coil continue to turn. The force is reversed by reversing the current in the coil. This is done by the **split-ring commutator**. You can see on the diagram that the d.c. supply is connected to the **brushes** – contacts which brush against the split-ring. The split ring rotates, and every half-turn the current in the coil changes direction.

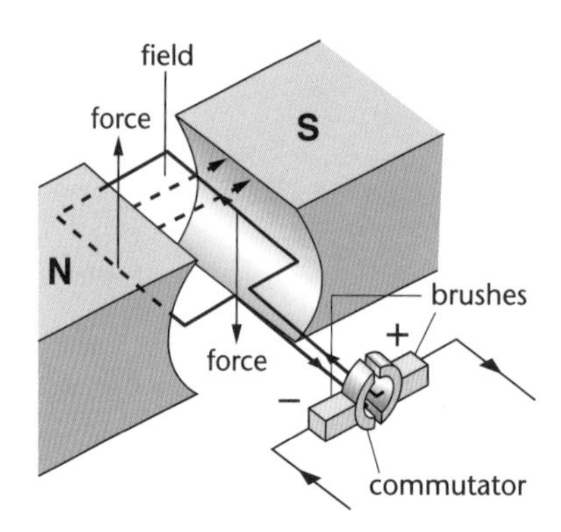

Fig. 13.11 A d.c. electric motor.

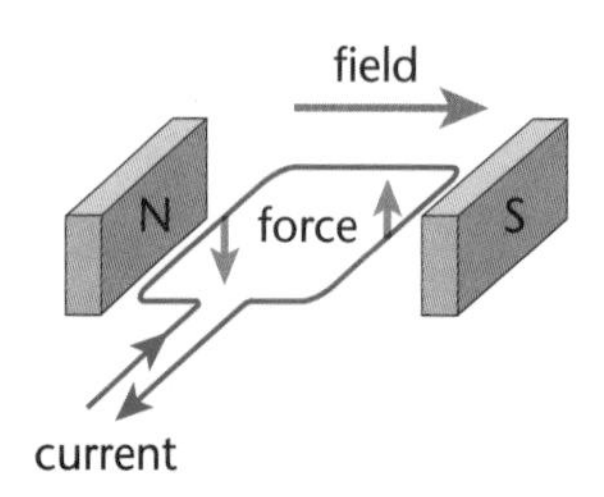

Fig 13.12 The forces on the coil in a d.c. electric motor

The speed of the motor can be increased by increasing:

- the size of the electric current
- the number of turns on the coil
- the strength of the magnetic field (stronger magnets and/or an iron core in the coil).

Practical electrical motors have curved pole pieces to produce a radial magnetic field, so that the force on the coil does not keep changing as it rotates.

In a radial field the field lines are towards the centre of a circle (along a radius).

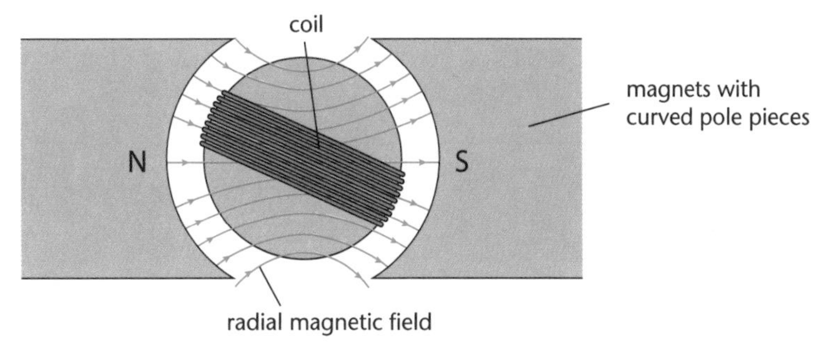

Fig. 13.13 Magnets with curved pole pieces give a radial field.

13.3 Diodes and capacitors

The diode

OCR B P6f

This follows from section 7.2 The diode.

The silicon diode

Silicon is a semi-conductor so that, unlike metals, it does not have free electrons. A **silicon diode** is made of a single piece of silicon. Each half of the silicon has a different impurity added. The impurities are elements, chosen so that:

- in one half the silicon has extra **electrons** that are free to move
- in the other half, the silicon has missing electrons – gaps called positive **holes**. The holes are free to move.

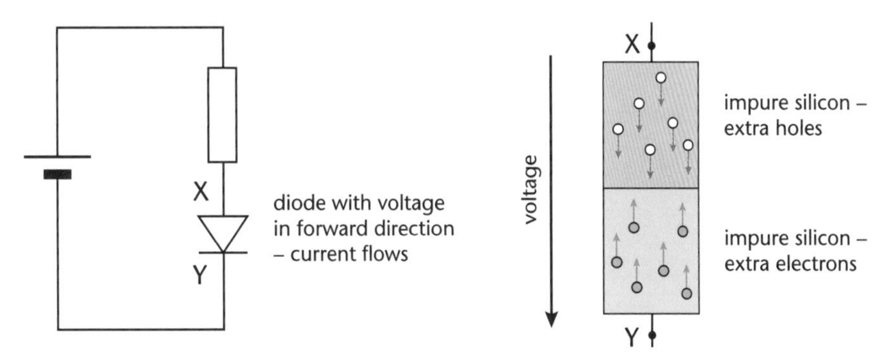

Fig. 13.14 A silicon diode.

When the voltage is in the forward direction, the current is made up of electrons moving one way, and holes moving in the opposite direction.

When the voltage is reversed, the electrons and the holes both try to move away from the junction and no current flows across the junction.

Half-wave rectification

When a diode is put in an a.c. circuit, the current will only flow in one direction. It will flow for half of the cycle and be blocked for the other half of the cycle. A graph of current against time will look like Fig. 13.15b. This is called **half-wave rectification**. The current now flows in only one direction, but half of the power has been wasted.

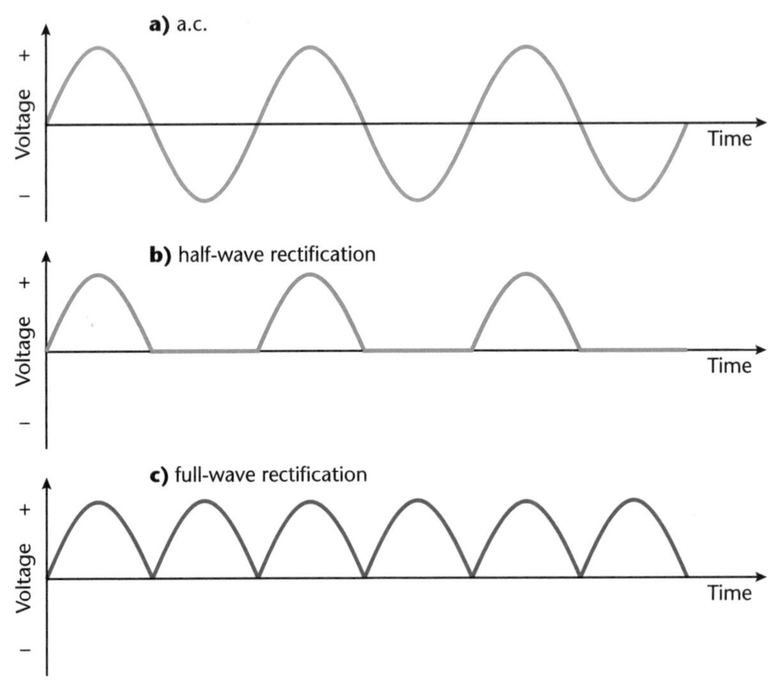

Fig. 13.15 The effect of diodes in a.c. circuits.

Full-wave rectification

Four diodes can be connected together in a **bridge circuit** to give **full-wave rectification**, as shown in Fig. 13.15c. In the bridge circuit shown in Fig. 13.16 the current for half the cycle flows along the red route from X through R to Y, and for the other half it flows along the blue route from Y through R to X.

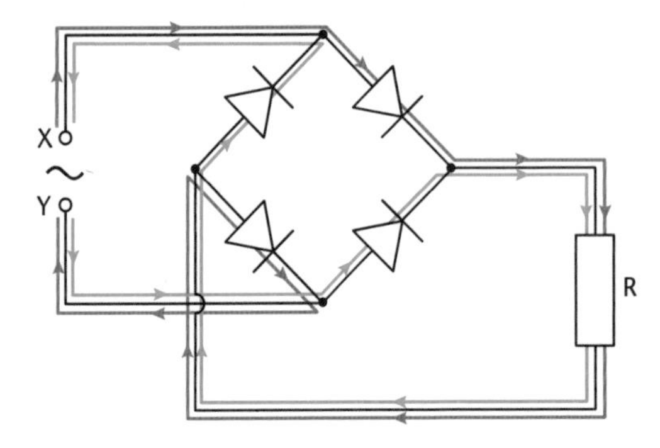

Fig. 13.16 A bridge circuit of four diodes. The current through R is always in the same direction.

The capacitor

A capacitor is a component that stores energy by separating electric charge. When a current flows in a circuit containing an uncharged capacitor, positive charge (+Q) builds up on one plate of the capacitor and an equal negative charge (–Q) builds up on the other plate. We say that the **capacitor stores charge** (Q). At the same time the p.d. across the capacitor increases. Fig. 13.17 shows a circuit for charging and discharging a capacitor. When the circuit is switched on in the charging circuit, the current has its maximum value. As the charge builds up, the p.d. across the capacitor increases in the opposite direction to the battery p.d. The current gets smaller and eventually, when the p.d. is equal and opposite to the battery p.d., the current is zero and the capacitor is fully charged. When the capacitor is discharged the current starts at its maximum value but as the charge flows the p.d. drops and the current gets smaller until eventually it reaches zero.

Electrolytic capacitors (circuit a) must always be connected with the + terminal to the + terminal of the supply. Other capacitors (circuit b) can be either way round.

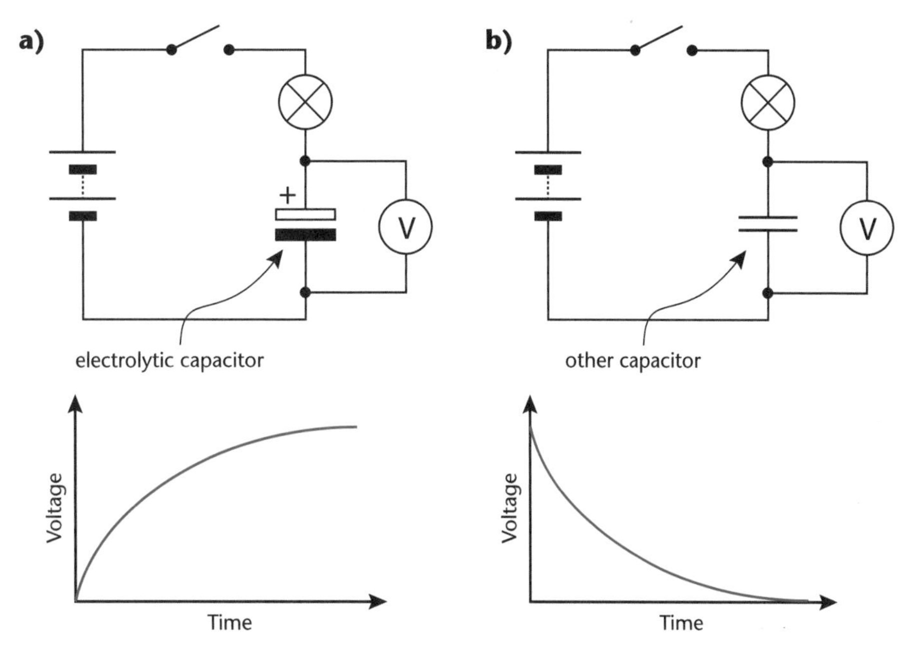

Fig. 13.17 (a) Charging and **(b)** discharging a capacitor.

In a d.c. circuit, once the capacitor is fully discharged, or charged, no more current flows.

A smoothing circuit

A lot of electronic appliances, for example computers, need a more constant voltage supply than a full-wave rectified supply (which keeps dropping to zero as shown in Fig 13.15c). A **smoothing circuit** uses a **capacitor** to stop the current from changing so much.

This is how it works:

1. The supply voltage increases, current flows through the resistor, and also charges the capacitor.

2. The supply voltage starts to reduce, so that the voltage across the capacitor is greater than the supply voltage. Charge starts to flow from the capacitor through the resistor.

3. The supply voltage increases again current flows from the supply through the resistor and recharges the capacitor.

The smoothed output has a small ripple, as shown in Fig. 13.18.

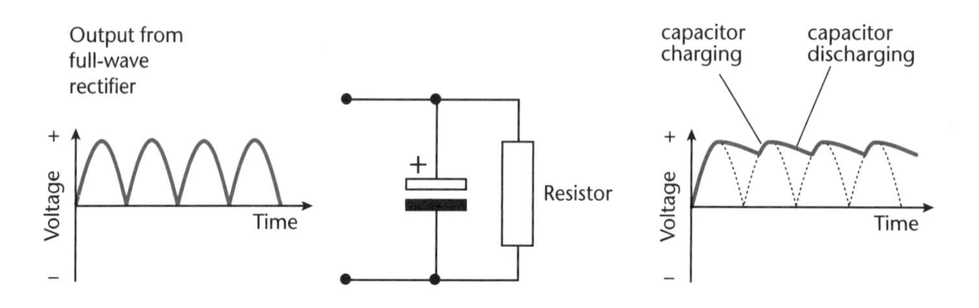

Fig. 13.18 A smoothing circuit using a capacitor.

13.4 Digital electronics

Logic gates

There are only two possible **states** in digital electronics, they are given different names and represented by two voltages in an electronic circuit, as shown in this table:

1st state	2nd state
on	off
high	low
about 5V	about 0V
1	0

A **logic gate** is a circuit with a **high** or **low output signal** that depends on the **input signal** (or signals).

These are the logic gates you need to know:

The NOT gate is also called an invertor.

The NOT gate, the OR gate and the AND gate

The symbols for these three logic gates, and their **truth tables**, are shown in Fig. 13.19. A truth table lists the output for all the possible combinations of inputs.

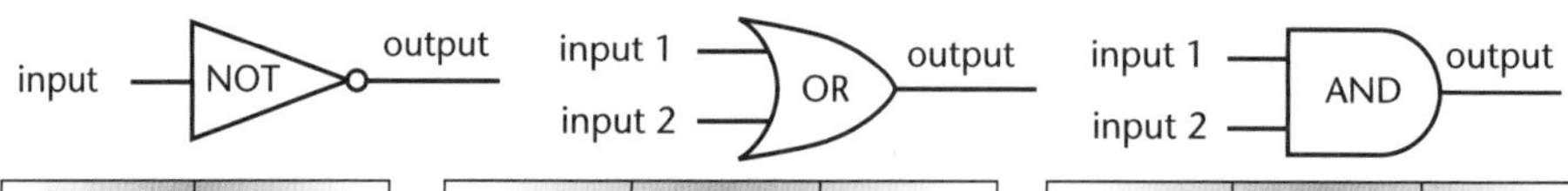

input	output
0	1
1	0

input 1	input 2	output
0	0	0
1	0	1
0	1	1
1	1	1

input 1	input 2	output
0	0	0
1	0	0
0	1	0
1	1	1

Fig. 13.19 The NOT gate, The OR gate and The AND gate and their truth tables.

The **NOT gate** has only one input. The output is always the state that the input is **not**, (the other state). The **OR gate** has a high output if either input 1 **or** input 2 (or both) are high. The **AND gate** has a high output only if input 1 **and** input 2 are high.

The NAND gate and the NOR gate

The **NAND gate** is the same as an AND gate followed by a NOT gate. The **NOR gate** is the same as an OR gate followed by a NOT gate. The symbols for these gates and their truth tables are shown in Fig. 13.20.

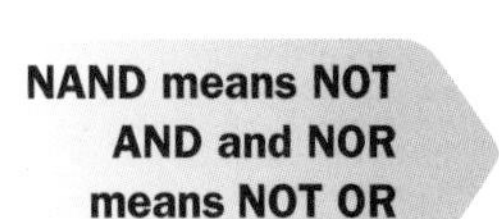

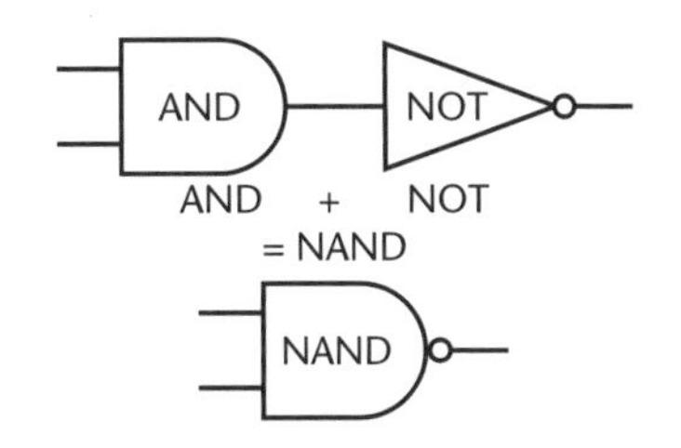

OR NOT
OR + NOT
= NOR
NOR

input 1	input 2	output
0	0	1
1	0	1
0	1	1
1	1	0

input 1	input 2	output
0	0	1
1	0	0
0	1	0
1	1	0

Fig. 13.20 The NAND gate, the NOR gate and their truth tables.

Generating an input signal

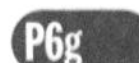

The p.d. across a thermistor in a potential divider circuit, as shown in Fig. 13.6a on page 205, can be used to generate a temperature dependent input p.d. for a logic gate. The threshold temperature at which the input switches from low (about 0 V across the thermistor) to high (about 5 V across the thermistor) can be adjusted by replacing R_1 with a variable resistor. In a similar way the p.d. across an LDR in a potential divider circuit, as shown in Fig. 13.6 b can be used to generate a light dependent input p.d. for a logic gate.

Fig. 13.21 shows a circuit that can be used to switch a heater on at night when the temperature falls below a certain level.

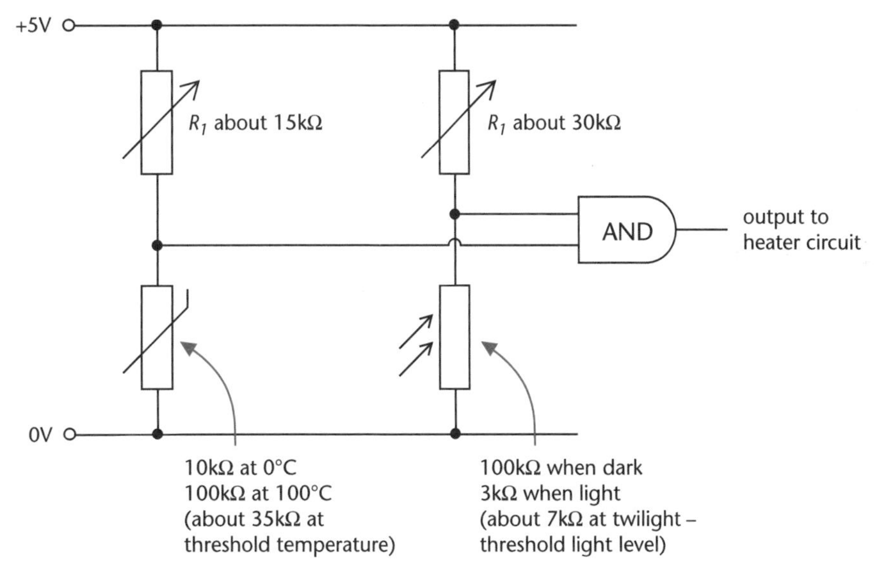

Fig. 13.21 A heater control circuit, variable resistor R_1 adjusts the threshold temperature and variable resistor R_2 adjusts the threshold light level.

Logic systems

OCR B P6h

A logic system can use a number of gates to combine input signals to give one output signal. Fig. 13.22 shows a simple 4 input combination lock.

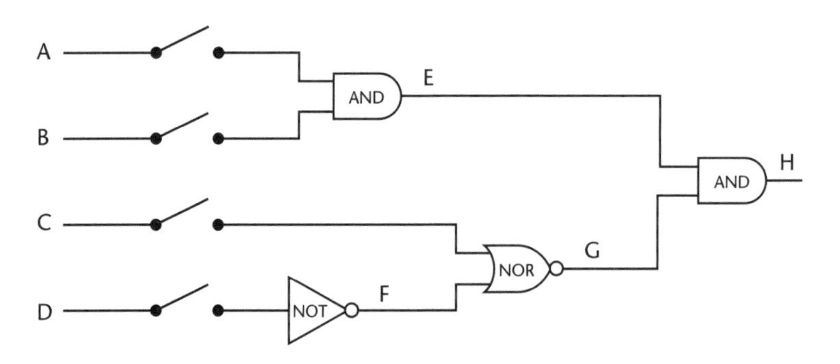

Fig 13.22 The circuit for a four input combination lock.

In the truth table, columns A, B, C and D are all the possible input combinations. Columns E, F and G are to help with working out the output, which is column H.

A	B	C	D	E	F	G	H
0	0	0	0	0	1	0	0
0	0	0	1	0	0	1	0
0	0	1	0	0	1	0	0
0	0	1	1	0	0	0	0
0	1	0	0	0	1	0	0
0	1	0	1	0	0	1	0
0	1	1	0	0	1	0	0
0	1	1	1	0	0	0	0
1	0	0	0	1	1	0	0
1	0	0	1	1	0	1	1
1	0	1	0	1	1	0	0
1	0	1	1	1	0	0	0
1	1	0	0	0	1	0	0
1	1	0	1	0	0	0	0
1	1	1	0	0	1	1	0
1	1	1	1	0	0	0	0

← This is the correct combination (row A=1, B=0, C=0, D=1)

The Bistable latch

OCR B P6h

Once a burglar alarm has been triggered, for example, by a burglar opening a door, it must keep ringing even if the door is closed. A **latch** circuit is used to keep the output at the same logic state even if the input has returned to its former state. A **bistable latch** has two output states that are both stable. These are shown in Fig. 13.23 a and b. Whether the latch is in state a or b depends on what has happened before – this circuit is a circuit with a memory.

When the **input S** goes **high** for **a moment**, as shown in Fig.13.23c, the output of the top NOR gate goes low. This low signal is input to the bottom NOR gate and the output – the latch output Q – goes high. The high signal is sent to the input of the top NOR gate, so that, even if input S goes low, the **latch output** will **stay high** – the circuit is latched. It will remain in this state **until input R** goes **high** for **a moment** as shown in Fig. 13.23d reversing the whole process so that the **latch output** will go low and **stay low**.

You cannot have both inputs high together because the output cannot be predicted.

Fig. 13.23 a and **b** the two stable states of a bistable NOR gate latch. The set input signal **c**, and the reset signal **d**.

Using the output of a logic gate

OCR B P6h

A light emitting diode (LED)

An LED can be used to show whether the output of a logic gate is high or low. The LED is used with a resistor in series so that a high voltage at the output will not cause a current that is too large for the LED.

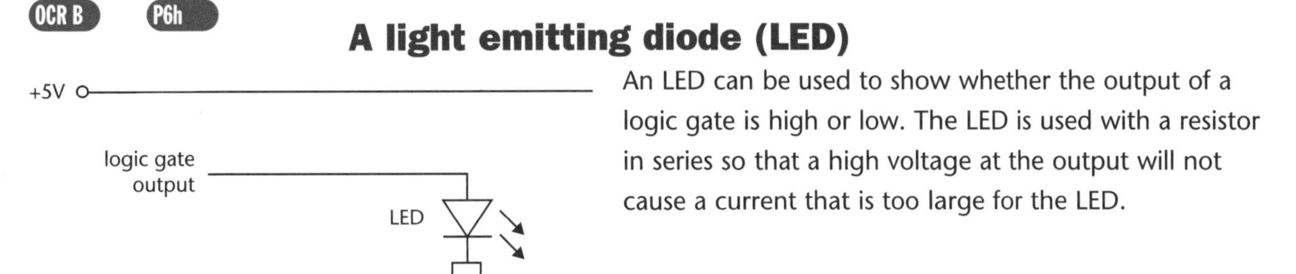

Fig. 13.24 An LED is used with a resistor in series.

The relay

A **relay** is a magnetic switch. It is used, for example, when we want the **low power** output logic circuit to switch a **large current** in a mains circuit. The relay also **isolates** the low voltage circuit from the high voltage mains circuit. Fig. 13.25 shows a circuit to switch on a heater using a relay. The small current in the logic circuit magnetises the relay coil, which attracts the iron arm of the switch in the second circuit, closing the switch so that the larger current flows. The relay has two connections for the coil, used by

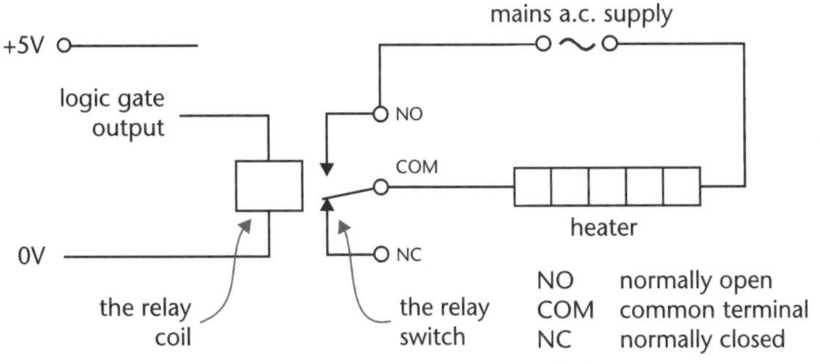

Fig. 13.25 A relay is used in a circuit to switch on a heater with the output from a logic gate.

the logic gate circuit, and three connections for the switch. One (labelled COM for common) is always connected to the mains circuit, but there is a choice of NO (normally open) or NC (normally closed), so that the relay can be used to switch on, or switch off, the circuit when the logic gate output is high.

13.5 Medical physics

Physics facts and medical treatment

EDEXCEL 360 P3.6

This section makes use of lots of different physics, applying it to medical diagnosis and treatment. There are now many non-invasive techniques available – especially for diagnosis. (Non-invasive means avoiding surgery.) The physics principles you need to know for the module are summarised in this table:

Physics principles applied in this section	**Book section**
Refraction of light	8.3 More about light and sound - refraction
Total internal reflection	3.5 Total internal reflection
Work, energy and power	6.4 Work, energy and power
Potential difference	7.2 Voltage (also called potential difference)
Frequency and time period Using an oscilloscope	8.1 Motion of particles in waves
Radiation intensity	4.1 The intensity of radiation
Nuclear equations	9.2 Different atoms 9.3 Radioactive emissions
Conservation of momentum	6.3 Momentum 11.1 Conservation of momentum
Positrons	10.3 Radioactive emissions 10.3 Stability of the nucleus
Effect of radiation on living things	1.2 Dangers of radiation
Using radiation to treat malignant tumours	9.3 uses of radioactive sources

Endoscopes

EDEXCEL 360 P3.6

Endoscopes can be used for non-invasive investigations.

Infrared radiation and light travel along glass **optical fibres** by being **totally internally reflected**, Fig. 13.26. The fibre is made with a central core that refracts more than the outer layer to ensure that total internal reflection keeps the light travelling along the fibre.

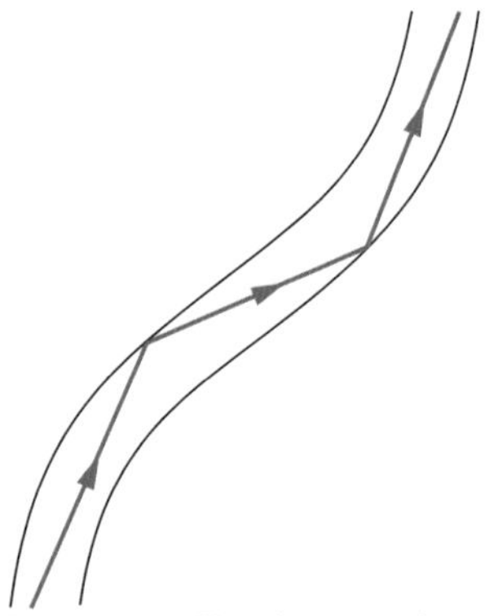

Fig. 13.26 Total internal reflection (TIR) in an optical fibre.

An **endoscope** is a bundle of optical fibres that can be used for viewing inside the body or for performing minor surgical operations. Optical fibres are flexible so the endoscope can be passed into the body through an opening, for example the mouth or nose. To view part of the body, for example the stomach, an endoscope has two sets of fibres. The first set **transmits light** to the stomach from a light source outside the body, and the second set transmits the **image** out of the body. In the second set, the fibres must stay lined up so that an image is formed. There is a lens system at the outside-the-body end of the fibres to transmit the image to a viewing screen or camera.

Endoscopes can have tools attached and be used for biopsies – removing a small piece of tissue to be tested in the laboratory – and for minor surgery or removal of foreign objects. The optical fibres can also transmit a **laser** light beam, which can be used to cut and heat tissue to stop bleeding.

Pulse oximetry

EDEXCEL 360 P3.6

Pulse oximetry is a method of measuring how much **oxygen** there is in a patient's blood. Blood that is oxygenated is bright red, while blood that does not have much oxygen is dark red or a bluer red, because the red blood cells absorb different wavelengths depending on how much oxygen there is in them. Blood in the veins always has less oxygen. The pulse oximeter is for measuring the oxygen in the blood in the **arteries**.

A pulse oximeter is used on a thin part of the body, usually a fingertip or earlobe, or the foot of a premature baby. A sensor (a photodiode) is placed on one side of the finger to detect a beam of **red light** (which you can see), from an LED (light emitting diode) and a beam of **infrared radiation** (which you can't see), from an IRED (infrared emitting diode). The ratio of the red light to the infrared radiation absorbed on passing through the finger gives a measure of the percentage of oxygen in the blood.

Light will also be absorbed by, for example, the blood in the veins, bone, skin, and muscle, so a pulse oximeter looks for the slight change in the ratio of radiation and light absorbed caused by the beat of the heart pushing blood through the arteries. This means pulse oximeters work best when there is a strong pulse in the part of the body being used.

Pulse oximeters are used to monitor oxygen levels in patients during intensive care, and in emergency treatment, as well as patients with heart problems.

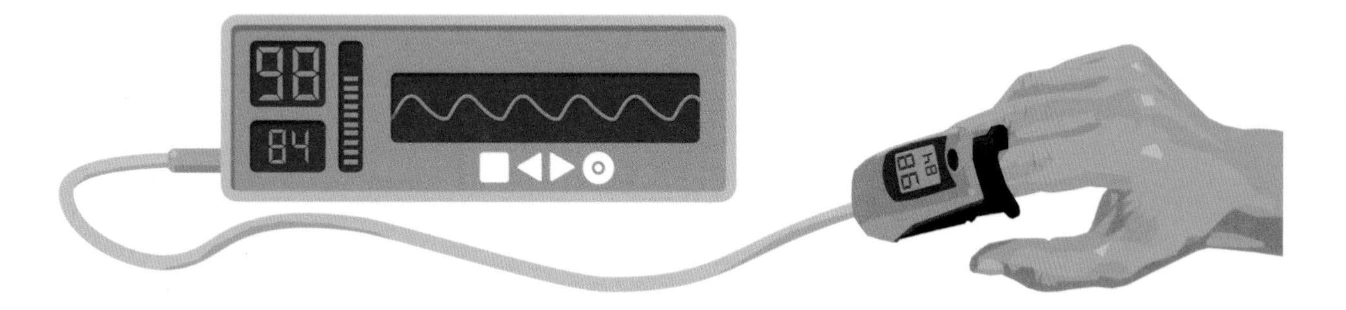

Fig. 13.27 A pulse oximeter in use.

Basal metabolic rate (BMR)

EDEXCEL 360 P3.6

Energy is needed for all the basic functions to keep the body alive.

For example:

- to transmit nerve impulses through the body
- to make new proteins needed for cell growth and repair
- for the muscular movements for vital functions such as breathing, blood circulation and digestion.

When the body is completely at rest, the rate at which energy is used is called the **basal metabolic rate**.

The rate varies from person to person, but for a young man it is about 85 W it is slightly less for a young woman, because she has less muscle tissue.

Potential differences in the body

EDEXCEL 360 P3.6

When muscle cells are not active they have a resting potential. This means that there is a **potential difference** generated across the cell membrane. When the muscle is activated the potential is momentarily reduced to zero, reversed, and then returned to the resting potential. This sequence of changes is called an **action potential**. The muscle contracts when the potential is reversed, and relaxes as it returns to the resting potential. The potential difference between cells can be measured, using electrodes placed on the body, and displayed on the screen of an oscilloscope (or on a computer), as a graph of potential difference against time. Skin is a poor conductor, so the site where the electrode is attached must be cleaned and some conductive paste rubbed in to the skin before the electrode is fixed in place.

The electrocardiogram (ECG)

EDEXCEL 360 P3.6

The **heart** is a pump controlled by **electrical signals**. One side of the heart takes in de-oxygenated blood from the body and pumps it to the lungs. The other takes in oxygenated blood from the lungs and pumps it round the body. There are one-way valves that only allow flow in the directions shown by the arrows in Fig. 13.28.

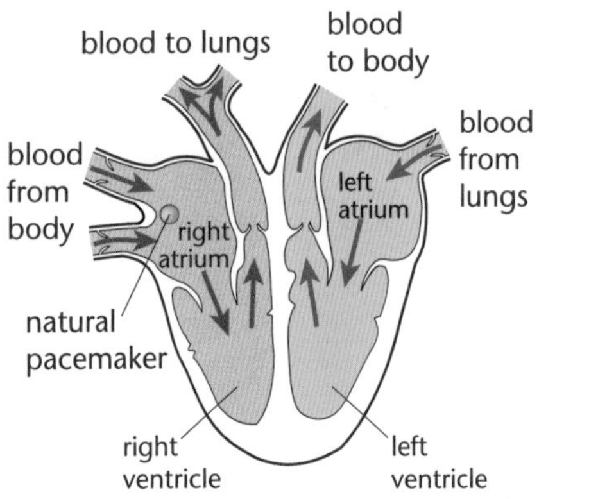

Fig. 13.28 The heart. The contraction starts in the area labelled natural pacemaker.

The electrical signals that control the heart are produced in the heart muscle itself (the heart will beat for a short time even when removed from the body). The rhythm is set by muscle fibres in the part of the heart labelled 'natural pacemaker' in Fig.13.28. The pulse starts in the pacemaker area and the **action potentials** spread through the **atria** making them contract and force blood into the ventricles. They then reach the **ventricles** making them contract and pump blood out of the heart.

Body fluids conduct electrical signals, so very small **potential differences**

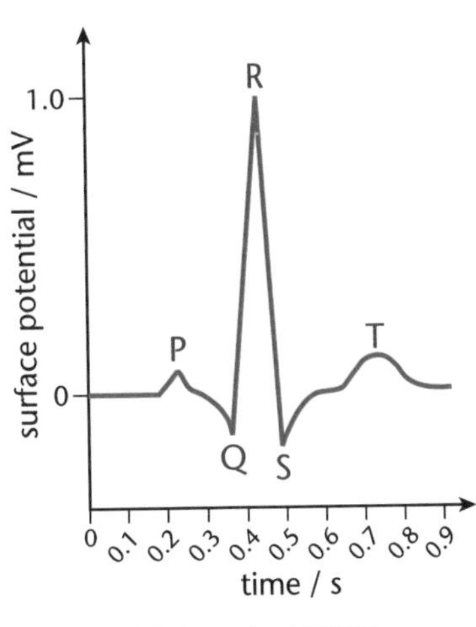

Fig. 13.29 A typical ECG

generated by the **action potentials** in the heart can be detected by electrodes on the surface of the skin. When these are amplified and recorded, they give a recording called an **electrocardiogram** (**ECG**). A typical ECG of one heartbeat is shown in Fig. 13.29.

Each beat is made up of:

- the P-wave, detected when the action potentials cause the atria to contract
- the QRS pulse, detected when the action potentials cause the ventricles to contract
- the T-wave, detected when the ventricles return to the resting potential, and the muscles relax.

The QRS pulse is so large that it hides the small wave that occurs when the atria return to the resting potential (and the atria relax).

Radioisotopes

EDEXCEL 360 P3.6

Some radioactive isotopes, also called **radioisotopes**, occur naturally, but many more are made in nuclear reactors. There a large number of different radioisotopes used in medicine, chosen for their particular properties.

Using neutrons

Some radioisotopes are produced by **bombardment** with neutrons. The neutrons produced during nuclear fission are very fast moving. To be captured by a nucleus, they need to be moving more slowly. When they have a speed that is about the same as the most probable speed of a gas molecule at room temperature they are called **thermal neutrons**.

A neutron has a mass number = 1 and a charge = 0 so is represented in nuclear equations by:

$^{1}_{0}n$

Example

Cobalt-60 **sources** are **gamma ray** emitters used to treat cancer. A cobalt-60 nucleus is produced when a cobalt-59 nucleus captures a thermal neutron:

$$^{59}_{27}Co + ^{1}_{0}n \rightarrow ^{60}_{27}Co$$

Using protons

Bombarding elements with protons usually produces radioactive isotopes that decay by positron emission (see section 10.3).

Example

An oxygen-18 nucleus is stable (although not as common as oxygen-16). When it is bombarded by protons, a proton is absorbed and a neutron emitted. The resulting nucleus is a fluorine-18 nucleus, which is radioactive, and decays by emitting a positron. The equation is:

$$^{18}_{8}O + ^{1}_{1}p \rightarrow ^{18}_{9}F + ^{1}_{0}n$$

Positron-electron annihilation

EDEXCEL 360 P3.6

The **positron** is the antiparticle of the electron. When an electron meets a positron they react and turn into **gamma rays**. After the reaction there is no particle with mass, so the particles are said to **annihilate** each other. Mass and energy are interchangeable, and the mass of the particles is converted to energy. **Conservation of energy** means that the energy of the positron and electron cannot disappear in the reaction. It is transferred by high-energy gamma rays.

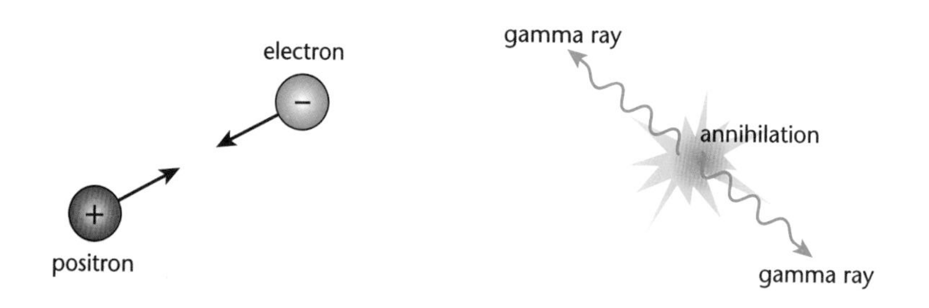

Fig. 13.30 An electron and a positron annihilate each other.

When a positron and electron annihilate each other:
Conservation of momentum means that two equal gamma rays are always produced that travel in opposite directions.

Einstein's famous equation $E = mc^2$ tells us that mass can be converted to energy and vice versa – you don't need to know the equation.

Positron emission tomography (PET) scanning

EDEXCEL 360 P3.6

Positron emission tomography (**PET**) is a three-dimensional medical imaging technique that is based on the emission of positrons.

The patient is injected with a medical tracer that contains a **radioisotope** that decays by emitting positrons. When the isotope decays, the positron travels only a few millimetres before meeting an electron. The positron and electron are annihilated and two **gamma rays** are produced which travel in opposite directions. The patient is placed in a scanner, which detects the gamma rays. Only pairs of gamma rays that arrive simultaneously are used in analysis, any other rays are ignored. From the position of the two gamma rays the point where they originated can be worked out, and a picture built up of the positron concentration.

The half-life of fluorine-18 is about 110 minutes, so the patient is not radioactive for a long time afterwards.

The tracer used depends on what the scan is for. A very commonly used radioisotope is fluorine-18. A nucleus of fluorine-18 is used to replace an oxygen nucleus in a sugar (glucose) molecule. When the radioactively tagged glucose is injected into the bloodstream it travels to parts of the body that use glucose for energy. Malignant **tumours** show up because the cells use glucose differently to normal cells. After injection, the patient waits for an hour for the glucose to spread through the body. The body is then scanned. This can take an hour.

The nucleus decays to an oxygen-18 nucleus, so the glucose does not change into a different compound – it is still glucose.

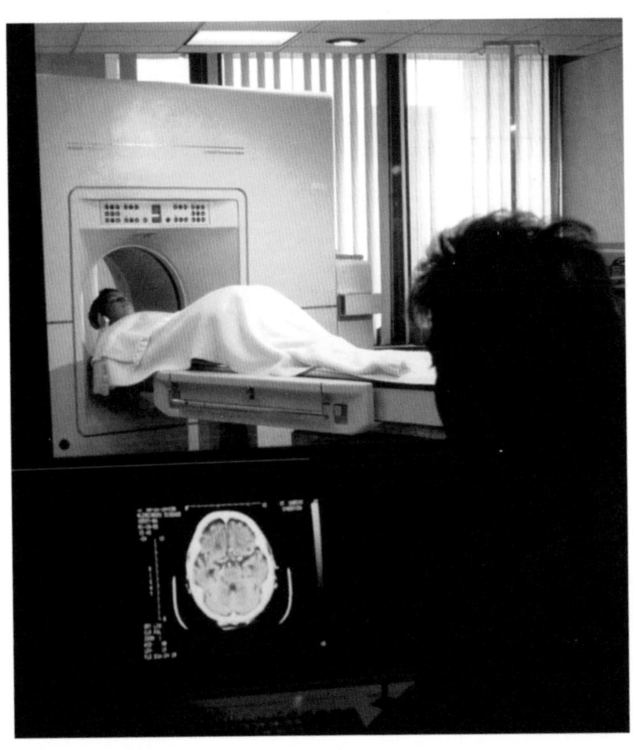

Fig. 13.31 These scans show different parts of the brain are active when seeing, hearing, thinking and speaking.

PET scans are useful for showing which parts of the brain are active when the patient is doing different things, because the brain uses glucose for energy. They can show the difference between Alzheimer's disease and other brain problems. They are used to diagnose and monitor cancer, particularly Hodgkin's disease, non-Hodgkin's lymphoma and lung cancer. They show whether cancer has spread to other parts of the body, and whether it has recurred after treatment.

Palliative care

EDEXCEL 360 P3.6

Radiation treatment does not always lead to a cure for the cancer or other condition being treated. Sometimes it can reduce suffering. This is called palliative care.

Examples are:

- to reduce bone pain when cancer has spread to the bones. Many patients find that radiation therapy reduces pain in this case
- to reduce headaches and difficulty thinking or seeing (focusing eyes) when cancer has spread in the brain.

HOW SCIENCE WORKS

Making decisions about new medical techniques

X-rays were discovered in 1895, in the same year they were used in an attempt to treat breast cancer. Similarly, when radioactive materials were discovered, scientists and doctors quickly began thinking about medical applications. In those days the radiation risks were not recognised and people were damaged by high radiation doses. Fortunately, enough data was collected that, once the dangers were realised, people had some idea of a safe dose – otherwise nuclear medicine might have been stopped at that time. There are claims that, by the year 2000, half of all cancer patients that survived owed their cure partly to radiotherapy.

So would the same thing happen with new techniques today?

Deciding whether it is safe and effective

Today there are many laws and official regulations to protect patients and staff and make sure that potentially dangerous techniques are properly tested.

The first tests are done on dead tissue – cells grown in the laboratory, or dead animals from a butcher's shop. (For example, transmission and absorption of laser light was tested on pieces of pig skin.) The next tests are done on human tissue. Some people leave their bodies to medical research. Scientists cannot just use tissue without the consent of the next of kin – although this has been done in the past. Some religions require the body to be disposed of in a particular way, or within a certain time.

Fortunately for medical research, there are many people who are not opposed to the use of their bodies after death. All these tests are ***in vitro* tests** (this means 'in glass' and refers to the fact that they are laboratory tests, usually using glass containers).

The next step is ***in vivo* tests**, tests using living tissue. The first tests are usually using animals, and people have lots of different views about animal testing – some are given below.

"Animals have rights – they should not be tested on."

"If they can find a cure for my boyfriend, I don't mind how many animals they test on."

"Animals don't have rights – but humans have a responsibility to treat them humanely."

"It should be allowed for life-saving medicines – but not cosmetics."

"Animals benefit from animal testing too."

"Testing that causes animals pain should not be allowed."

"My wife would have died without the new treatment – it was tested on animals."

"It's a few animals against thousands of people."

HOW SCIENCE WORKS

There are regulations to ensure that the animals are treated as humanely as possible.

The next step is to set up **clinical trials** to test the technique on people. About 10–50 volunteers might be selected. They would be carefully monitored to check for side-effects. If the technique appears to be safe, then a larger trial will be tried, and if that is successful a large trial of several thousand patients will be tried.

For a scanning technique or a tracer, the results can be seen on the scans, so it is not quite the same as testing a drug. However, there may still be instances where doctors use a blind trial. In a blind trial, some patients are given the treatment and others are given a **placebo** – something that looks like the treatment but does nothing – as a control. This is so that the patients do not know whether they have had the treatment or not, so that their expectations do not affect the results. In a **double-blind trial** the doctor does not know which patients have received the treatment. It may be difficult to decide whether to take part in a trial (some people's opinions are shown below).

"Suppose I have the placebo – I won't be getting any treatment."

"Suppose I have the drug and it has side-effects"

"I couldn't get on the trial – so I don't have the chance to try the drug."

Sometimes the results are so conclusive early on that the trial is stopped and the treatment is used for everyone – or withdrawn. People on the trial must be given all the information, and be told they can leave at any time. They sign a consent form to say they understand this.

1. (a) Explain why some people feel that testing on animals should be allowed.
   (b) Explain why some people feel that testing on animals should not be allowed. **[2]**
2. Write a short introduction to a leaflet for patients who are being asked to take part in a double-blind clinical trial of a new tracer for PET scanning (the control group will receive the tracer that is currently being used). In your introduction explain:
   - what a double-blind clinical trial is
   - what their options are. **[2]**

Exam practice questions

1. Which of these transformers transforms 230 V a.c. mains down to 12 V?

	Number of turns	
	Primary coil	**Secondary coil**
A	12	230
B	48	920
C	460	12
D	920	48

[?]

2.

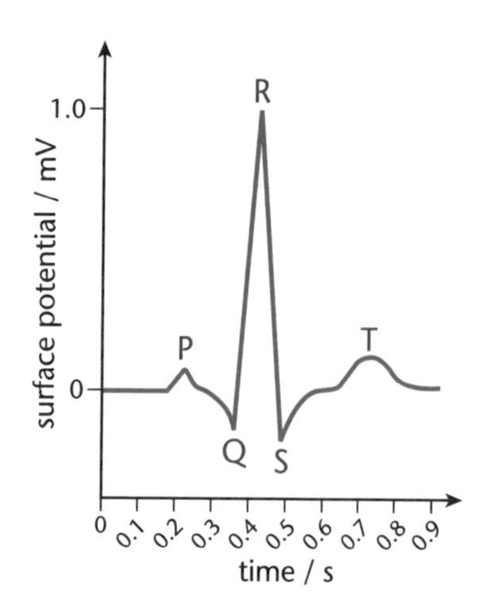

The T- wave occurs when:

A the atria contract

B the atria relax

C the ventricles contract

D the ventricles relax

[1]

3. This circuit has three input signals and one output signal. Which column of the truth table, A, B, C or D shows the output at X?

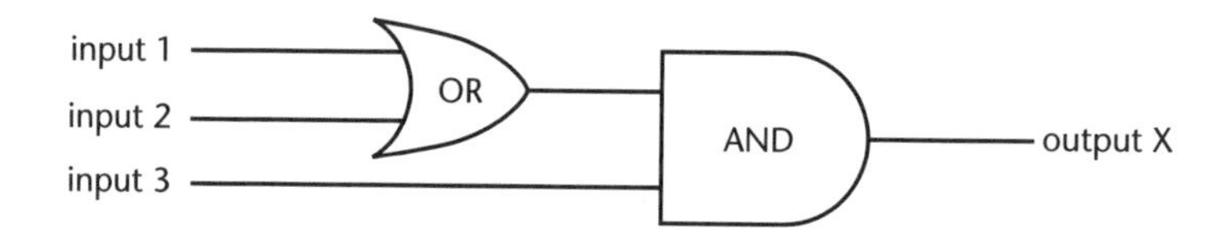

Input 1	**Input 2**	**Input 3**	**A**	**B**	**C**	**D**
0	0	0	0	0	0	0
0	1	0	1	0	0	0
0	0	1	0	0	0	0
0	1	1	1	1	1	1
1	0	0	1	0	1	0
1	1	0	1	0	1	1
1	0	1	1	1	1	0
1	1	1	1	1	1	1

[1]

Exam practice questions

4. Match the component to the correct circuit symbol.

Component

thermistor

capacitor

light emitting diode (LED)

Circuit symbol

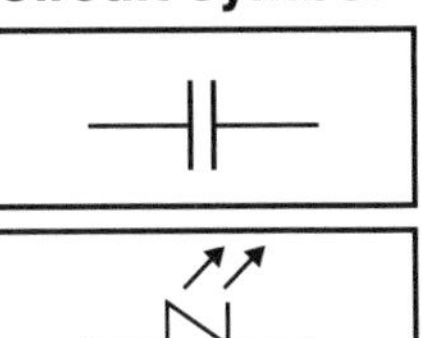

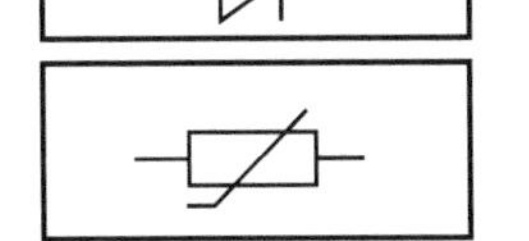

[1]

5. These statements are about an electric motor. Write **T** for the **True** statements and **F** for the **False** statements.

(a) The force on the side of the coil is greatest when the direction of the electric current is parallel to the direction of the magnetic field.

(b) The split-ring commutator is used to reverse the direction of the current in the coil every half turn.

(c) Increasing the number of turns on the coil increases the speed of rotation

(d) The brushes are used to keep the coil clean.

(e) Reversing both the direction of the electric current and the direction of the magnetic field reverses the direction of the coil rotation.

(f) The curved pole pieces and the iron core give a radial field to keep the speed of rotation constant. **[3]**

6.

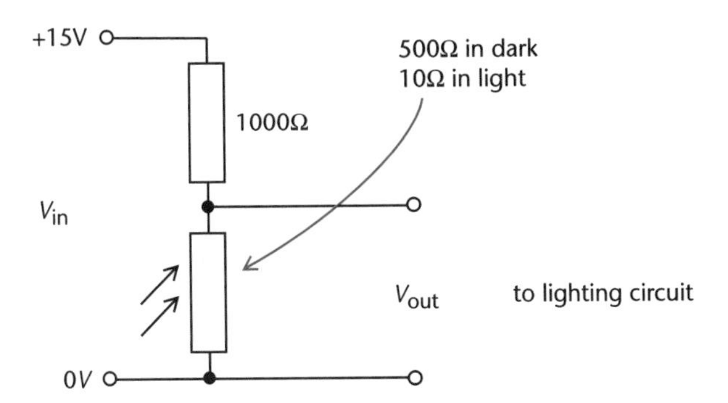

Fill in the gaps to complete the sentences. (Use words once, more than once, or not at all.) Choose from:

10 Ω   500 Ω   1000 Ω   0 V   0.15 V   5 V   10 V   15 V
dark   LDR   LED   light   resistor   thermistor   variable

When it is dark the resistance of the LDR is ____1____ ohms and V_{out} = ___2___ V. When it is light the resistance of the LDR is ____3____ ohms and V_{out} = ___4___ V. This circuit can be used to switch a light on when it gets ____5____. The light level at which the light switches on or off can be altered by replacing the ____6____ ____7____ with a ____8____ ____9____. **[8]**

Exam practice questions

7. An adaptor for a computer is required to change the 230 V mains a.c. electricity supply to a 12V d.c. supply.

(a) What is the name of the equipment needed to change the 230 V a.c. supply voltage to a 12 V a.c. supply?

(b) If there are 230 turns in the primary coil, how many are needed in the secondary coil?

(c) What is the name of the component that can be used for half-wave rectification - to change a.c. to d.c.

(d) How many of these components would be needed for full-wave rectification?

(e) Sketch a graph of the output voltage against time for the full wave rectified output voltage.

(f) Explain what else needs to be done to make the supply voltage suitable for a computer and suggest how this could be done. **[8]**

8. A patient is given a tracer that is a gamma emitter and after a short time a gamma camera is used to produce a scan of the patient. A second patient is given a PET scan. PET stands for Positron emission tomography.

(a) Write down one similarity between the two techniques.

(b) Write down one difference between the two techniques.

(c) Explain how radioisotopes that are positron emitters can be artificially produced.

(d) How is the positron emitter carried to the part of the body that is under investigation.

(e) Give one example of when a PET scan is used. **[6]**

9.

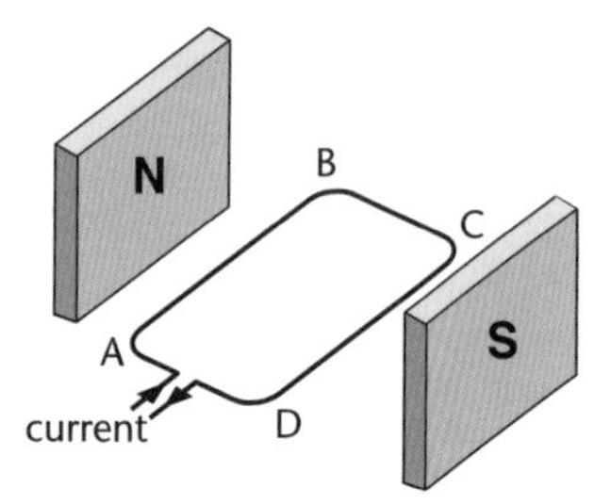

This diagram shows an a.c. generator when the coil, ABCD is horizontal, and the current flows from A to B then to C and then to D.

(a) How is the current different when the coil reaches:

(i) the vertical position with AB at the top?

(ii) the horizontal position with DC closest to the north pole of the magnet?

(iii) the vertical position with DC at the top?

(b) State three ways that the current in the coil can be increased.

(c) What extra components are needed to connect the coil to an external circuit?

(d) The coil is rotating so that it makes one complete turn every 25 milliseconds. What is the frequency of the a.c. generated? **[9]**

Rearranging formulae

If you are asked to write down a formula or equation, do not write down only the triangle – or you will get no marks.

The formulae that you need in the exam are written on the exam paper. OCR A and OCR B have a sheet labelled useful 'relationships' or 'formulae'. Edexcel 360 has each formula written in a box as part of the question.

You may need to rearrange a formula. Some students find using this triangle method useful.

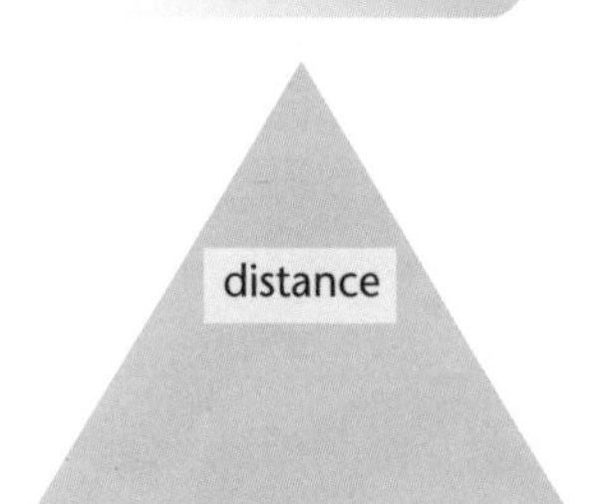

Example 1 speed = $\frac{\text{distance}}{\text{time}}$

To calculate distance:

1. Write the formula into a triangle, so that distance is 'over' time. This means putting distance at the top, time can go in either of the other corners.
2. Cover the word distance with your finger and look at the position of speed and time. They are side by side, so distance = speed x time.
3. To find the time, cover time with your finger and distance is 'over' speed, so time = $\frac{\text{distance}}{\text{speed}}$

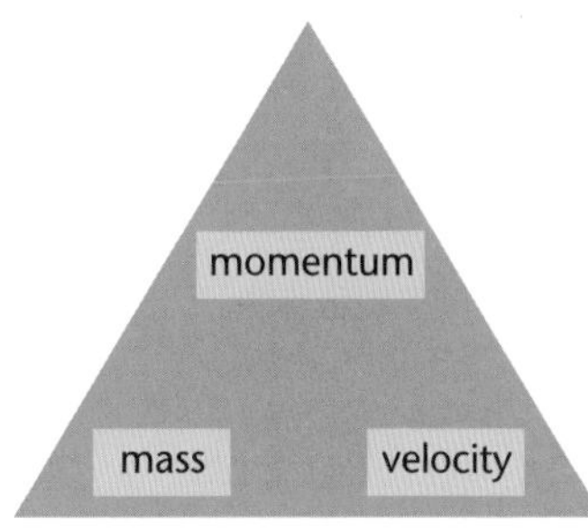

Example 2 momentum = mass x velocity

1. Write the formulae into a triangle so that mass and velocity are side by side. This means they must go in the two bottom corners of the triangle.
2. Covering mass gives mass = $\frac{\text{momentum}}{\text{velocity}}$
3. Covering velocity gives velocity = $\frac{\text{momentum}}{\text{mass}}$

	OCRA	OCRB	AQA	EDEXCEL
Energy = mass x specific heat capacity x temperature change		P1a		
Energy = mass x specific latent heat		P1a		
speed = $\frac{\text{distance}}{\text{time}}$	P4.1	P3a		
velocity = $\frac{\text{displacement}}{\text{time}}$ $v = \frac{s}{t}$	P4.1	P5b		P2.9
acceleration = $\frac{\text{change in velocity}}{\text{time taken for change}}$ $a = \frac{(v-u)}{t}$		P3b	P2.12.1	P2.9
weight = mass x gravitational field strength $W = m \times g$		P3h	P2.12.2	P1b12
Resultant force = mass x acceleration $F = m \times a$		P3c	P2.12.2	P2.9
Kinetic energy = $\frac{1}{2}$ mass x speed ^{2} KE = $\frac{1}{2} m v^2$	P4.4	P3e	P2.12.3	P2.10
Change in gravitational potential energy = mass x gravitational field strength x height difference PE = $m \times g \times h$	P4.4	P3h		P2.10
momentum = mass x velocity	P4.3	P5d	P2.12.4	P2.9
force = $\frac{\text{change in momentum}}{\text{time taken for change}}$	P4.3	P5d	P2.12.4	
potential difference = current x resistance resistance = $\frac{\text{voltage}}{\text{current}}$ $V = I \times R$	P5.2	P4c, P6a	P2.12.6	P1a9
power = $\frac{\text{energy transformed}}{\text{time}}$	P5.5		P1.11.3 P2.12.8	
Power = $\frac{\text{work done}}{\text{time taken}}$ $P = \frac{W}{t}$		P3d		P2.10, P3.6
Energy transferred (or transformed) = work done	P4.4		P2.12.3	P2.10
Work done = force x distance moved in direction of force $W = F \times s$	P4.4	P3d	P2.12.3	P2.10, P3.6
power = current x potential difference or power = current x voltage	P5.5	P2c	P2.12.8	P1a10
energy transferred (or transformed) = potential difference x charge			P2.12.8	
charge = current x time			P2.12.8	
Electrical energy = voltage x current x time $E = V \times I \times t$	P5.5			P2.10

	OCRA	OCRB	AQA	EDEXCEL
Cost = power x time x cost of 1kWh				P1a10
For a transformer the ratio of the voltages is the same as the ratio of turns in the coils; $\frac{V_p}{V_s} = \frac{N_p}{N_s}$	P5.4	P6e	P3.13.9	
Efficiency = energy usefully transferred/total energy supplied x 100	P5.5	P1b, P2b	P1.11.2	P1a10
Wave speed = frequency x wavelength	P6.1	P1g	P1.11.5	
Power = $\frac{1}{\text{focal length}}$	P7.2			
Magnification = $\frac{\text{focal length of objective lens}}{\text{focal length of eyepiece lens}}$	P7.2			
Speed of recession = Hubble constant x distance	P7.3			
Moment = force x perpendicular distance from the line of action of the force to the axis of rotation			P3.13.1	
$v = u + at$		P5b		
$s = \frac{(u + v)}{2} t$		P5b		
$v^2 = u^2 + 2as$		P5b		
$s = ut + \frac{1}{2} a t^2$		P5b		
Refractive index = $\frac{\text{speed of light in a vacuum}}{\text{speed of light in a medium}}$		P5g		
$n = \frac{\sin i}{\sin r}$		P5g		
$\sin c = \frac{n_r}{n_i}$		P5g		
Magnification = $\frac{\text{image size}}{\text{object size}}$		P5h	P3.13.4	
$V_{out} = V_{in} \times \frac{R_1}{(R_1 + R_2)}$		P6b		
$V_p I_p = V_s I_s$		P6e		
$\frac{P_1 V_1}{T_1} = \frac{P_2 V_2}{T_2}$				P3.5
KE = e x V				P3.5
Frequency = $\frac{1}{\text{period}}$			P2.12.7	P3.6
Intensity = $\frac{\text{power of incident radiation}}{\text{area}}$				P3.6

Answers

Chapter 1

How Science Works

1. **(a)** E **(b)** A
2. No because the range overlaps.
3. **(a)** 140 track counts **(b)** 70 track counts
4. **(a)** true **(b)** false **(c)** false **(d)** false **(e)** true **(f)** true **(g)** true

Exam practice questions

1. **D**
2. **D**
3. **C**
4. Nuclei, unstable
5. **1** Fuel rods, **2** nucleus, **3** neutron, **4** neutrons, **5** fission, **6** chain, **7** control rods, **8** neutrons, **9** coolant
6. 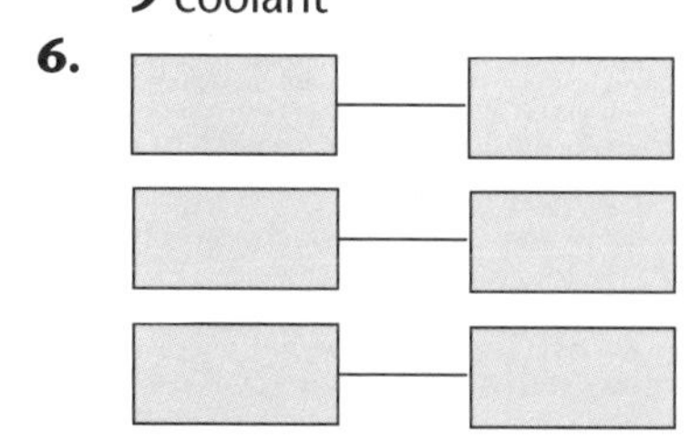
7. **(a)** as low as reasonably achievable
   (b) You should not take a risk if you are not sure of the consequences.
   It can be summarised as 'better safe than sorry'
8. **(a)** Beta radiation
   (b) No alpha radiation would get through the paper. All the gamma radiation would get through the paper. Only beta radiation intensity would vary with thickness
9. **(a)** One-sixteenth
   (b) **(i)** It would decay away before reaching the organ and being recorded by the gamma camera
   (ii) The patient would have radioactive material in him/her for days
10. **(a)** A nucleus splits in two
    (b) A million times more in a nuclear reaction

Chapter 2

How Science Works

1. **(a)** Joanna **(b)** David **(c)** Ann **(d)** Gemma
2. Q1 and Q3

Exam practice questions

1. **C**
2. **C**
3. **B**
4. **A**
5. 8400, 0.5, more than
6. **1** Conduction, **2** convection, **3** radiation (any order), **4** convection currents, **5** air gaps, **6** conduction, **7** walls, **8** radiation.
7. **(a)** T **(b)** F **(c)** F **(d)** T **(e)** T **(f)** F
8. **BDAC**
9. **(a)** 58%
   (b) Burns gas not coal, uses hot gas and steam, coal uses steam only, *or* more efficient (58% compared to about 40%)
   (c) Produces carbon dioxide which causes global warming *or* produces pollution, e.g. sulfur dioxide *or* is a fossil fuel so will run out
10. **(a)** The amount of wind (or wind speed)
    (b) C
    (c) **(i)** The more powerful appliance used at that time of day – e.g. cooker
    (ii) At night everything might be switched off
    (iii) If the wind drops below the maximum the turbine will produce less.
    Will not cope with peaks in demand
    (d) Can use batteries when there is no wind or to give a more constant supply
    (e) **(i)** 500 **(ii)** 1500
    (f) The time for the turbine to save the amount of money that it cost to buy and install
11. **(a)** 1748 units
    (b) £174.80
    (c) 6
    (d) 60p
    (e) the lamp
    (f) 5p

Chapter 3

How Science Works

1. **4**
2. **C**
3. **(a)** There was an increase in small earthquakes before a large one.
   (b) Some people sense/some animals behave strangely where there is going to be an earthquake.

Exam practice questions

1. **B**
2. **B**
3. **A**
4. **1** disturbance, **2** medium, **3** energy, **4** matter, **5** vibrates
5. **1** speed, **2** wavelength, **3** frequency, **4** direction

6.

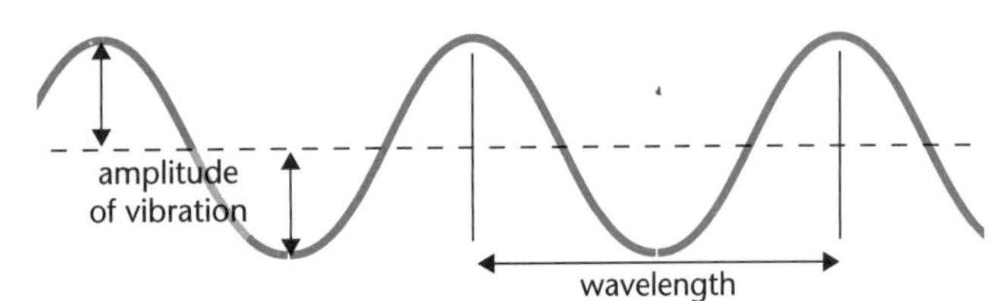

7. **(a)** 10s
(b) 0.1 Hz
(c) 20 m/s

Chapter 4

How Science Works

1. **(a)** Same numbers at the start and end for each group (phone users and non-users). Very small sample – results could be just random chance
(b) Bigger sample will give more meaningful results – but by the end of the trial sample not using phone is small; 0–2-year-olds cannot use phones
(c) If the 200 who start to use a phone are removed there is a sample of 800, 400 who use a phone and 400 who don't
(d) Not really – as shown by the Sun cancer case it could take longer for effects to be seen
(e) Texting does not expose the brain to so many microwaves, so could make using a mobile phone look safer than it is. A good study would need to know more about the actual phone use

2. **(a)** answer between 1.9 and 2.3 per 100 000 population
(b) answer between 2.8 and 3.2 per 100 000 population

Exam practice questions

1. **D**
2. **B**
3. **D**
4. **1** infrared, **2** optical fibre, **3** noise, **4** analogue, **5** digital, **6** noise, **7** regenerated
5. **(a)** T **(b)** T **(c)** F **(d)** F **(e)** F **(f)** T
6. **1** vibrate, **2** microwaves, **3** metal, **4** reflects, **5** microwaves, **6** grid, **7** reflect, **8** transparent.
7.

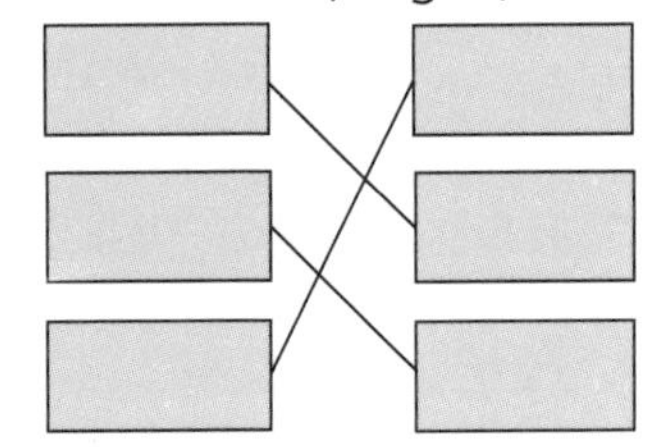

8. **(a)** Communication using electromagnetic waves without wires
(b) Two of: We can receive phone calls and e-mail 24 hours a day; no wiring is needed to connect laptops to the Internet, no wiring is needed to connect phones or radio communications; is portable and convenient

Chapter 5

How Science Works

1. **(a)** The galaxies are moving away from each other
The cosmic microwave background radiation
(b) It might slowly expand and then contract, there might be another Big Bang (oscillating Universe). It might continue expanding forever. It might stop expanding (steady state Universe)
(c) The amount of mass in the Universe
(d) No, (not as far as we know) because there is no evidence as nothing exists from before the Big Bang.

2. C, A, E, B, D

Exam practice questions

1. Earth, Sun, Solar System, Milky Way galaxy, Universe
2. **C**
3. **D**
4. **1** nebula, **2** contracts, **3** protostar, **4** nuclear, **5** fusion, **6** hydrogen, **7** helium, **8** main sequence, **9** hydrogen, **10** red giant, **11** planetary nebula, **12** white dwarf, **13** black dwarf
5. **(a)** F **(b)** F **(c)** T **(d)** T **(e)** T **(f)** T

Chapter 6

How Science Works

1. mass/the rate that time passes
2. B
3. Because the time shifts measured were all less than 300nm; so the results would not have been significant/would have been within the range of the measurements
4. It took this long to develop a stable and accurate enough clock.

Exam practice questions

1. **(a)** **(i)** D
(ii) C
(b) D
(c) C
2. 1 pair, 2* equal, 3* opposite, 4* action, 5* reaction, 6 foot, 7 backward, 8 ground, 9 ground 10 forward, 11 foot, 12 friction, *2 and 3 can be reversed, 4 and 5 can be reversed
3. (D) B C E A F

4. lines joining A2 B1 C4 D3
5. **(a)** 800 x 10 x 80 = 640 000J
   (b) gain in KE = loss in PE = 640 000J
   (c) 640 000 = $\frac{1}{2}$ mv^2 = $\frac{1}{2}$ x 80 v^2,
   v = √ 1600 m/s = 40 m/s
   (d) 25 000 ÷ 10 = 2 500 N
6. **(a)** steady speed, resultant = 0, so driving force = 5000 N
   (b) $\frac{1}{2}$ x 3000 x (30)^2 J = 1 350 000 N
   (c) momentum = 3000 x 30 = 90 000 kg m/s
   OR acceleration = (0–30) ÷ 5 = -6 m/s^2 (- for deceleration)
   (d) change in momentum ÷ time = –90 000 ÷ 5 = –18 000 N
   OR mass x acceleration = 3000 x (-6) = –18 000 N
7. **(a)** 36 m
   (b) **(i)** drinking, tiredness etc (see page 96)
   (ii) icy road, poor tyres etc (see page 96)
   (c) $\frac{1}{2}$ x 60 x (20)^2 = 30 x 400 = 12 000 J
   (d) seatbelt stretches
   (e) reduces the stopping force on the body
   (f) The person feels safer, that the risk is less, so it is OK to drive faster
   (g) Road travel
   (h) Because she travels by road everyday this is a risk she is used to, but travelling by rail is an unfamiliar risk and people over-estimate unfamiliar risks

Chapter 7

How Science Works

1. A
2. 2.0 V 8.8 mA
3. 2.5 V
4. **(a)** 0.9 mA
   (b) 17.8 mA

Exam practice questions

1. B
2. D
3. B
4.

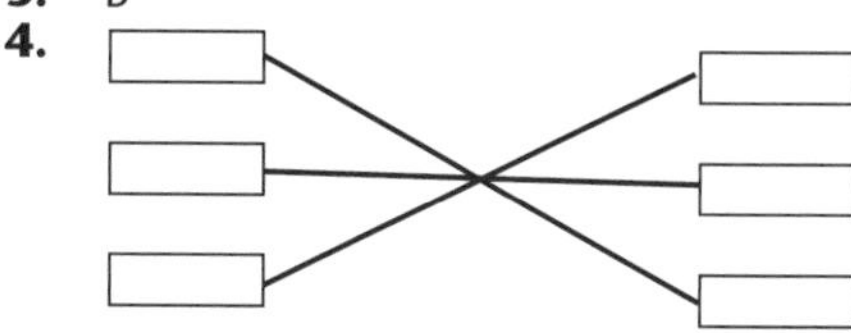

5. B E D C A
6. 1. chest 2. electrical 3. contract 4. charge 5. shock
7. **(a)** the case is a non-conductor
   (b) **(i)** current flows normally
   (ii) fuse melts and breaks circuit
   (c) will not melt until 13A flows, much higher current than needed
8. **(a)** the droplets will be charged and repel each other
   (b) droplets will be attracted to car body
   (c) even coverage all over – in shadows etc; less waste paint
9. **(a)** X is the ammeter Ⓐ Y is the voltmeter Ⓥ
   (b) 1050 mV
   (c) **(i)** brighter
   (ii) off
   (iii) dimmer
   (d) An a.c. power supply produces a current that keeps changing direction; d.c. is always in the same direction.
   (e) The current will only flow in the circuit when the voltage is in the forward direction. When the voltage is in the reverse direction the current will be zero, so the diode will flash on and off 50 times a second (so fast you can't tell).

Chapter 8

How Science Works

1. ADFBHGCE
2. Isaac Newton said light was not a wave and scientists thought he was such a great scientist he must be right
3. no, because light shows wave behaviour; diffraction/interference/an example described such as Young's experiment

Answers to questions

1. A
2. B
3. D
4. true, false, true, false
5. modulate, carrier, transmitter, aerial, signal, demodulated
6.

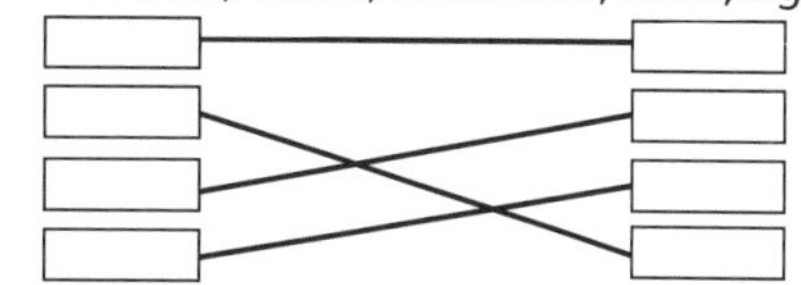

7. **(a)** sound with frequency higher than the upper threshold of human hearing
   (b) Some is reflected from each boundary between tissue layers. The reflections are used to build up an image
   (c) breaking up stones e.g. kidney stones
8. **(a)** A = compression B = rarefaction
   C = wavelength D = longitudinal
   (b) particles move up and down but they do not move along

Chapter 9

How Science Works

1. **(a)** No because radon is an alpha emitter and drinking it would expose you to a risk of cancer. There is no health benefit.
   (b) No, because there are safer ways to get a point to a brush, and the radium was a health risk.
   (c) Yes because she would have died many years earlier from cancer if she had not been treated
   (d) Yes because the cancer can be slowed or cured by the radioactive treatment, prolonging or improving the quality of life
2. There are risks to health and no benefits and it would contaminate people's homes
3. **(a)** The risk from being exposed to normal background amounts to 15 days lost on average, whereas the risk from smoking amounts to 6 years lost on average, so smoking is much more dangerous.
   (b) people are more worried about unfamiliar things and overestimate the risk.

Exam practice questions

1. B
2. B
3. B
4. C D F B A E G
5. **(a)** background radiation is the level of radioactive emissions that is all around us all the time.
   (b) one of, for example: cosmic rays, the Sun, rocks, leaks from radioactive waste.
6. **(a)** T **(b)** F **(c)** T **(d)** F **(e)** T **(f)** T
7. alpha, gold, most, empty, space, small, positive
8. **(a)** sodium-24 has one extra neutron
   (b) 15 hours
   (c) 20 counts per minute
   (d) **(i)** no because it would decay too fast and there would be no radioactivity left after a few weeks/because it is a gamma emitter and people who came close to it would be irradiated

(ii) Yes because it is a gamma emitter; so the radiation from the pipe would reach the surface/because it has a short enough half-life not to contaminate the site for a long time

Chapter 10

How Science Works

1. Economic – can pool money; Collaboration – can share expertise and ideas
2. advantage: jobs,
   disadvantage: dependent on the project continuing/ cost of housing etc may rise
3. **(a)** examples are: people who can support each other with information e.g. cancer sufferers/ students can get information / sellers or buyers of specialist or unusual items who would not meet otherwise
   (b) examples are: shops that lose business to the internet/ books or magazines that lose sales/ copyright owners whose material is distributed free (e.g. music.), victims of crimes that have been encouraged by the internet (e.g. fraud)
4. You can answer yes or no to this question. The marks are for a good justification of your choice e.g. yes because the number of people who benefit is larger than the number who don't.
   Or: no because the people who benefit could manage without the internet as they did before, but the people who have been victims of criminals they met through the internet have in some cases been murdered and this cannot be justified.

Answers to questions

1. D
2. A
3. C
4. T F F T F
5. 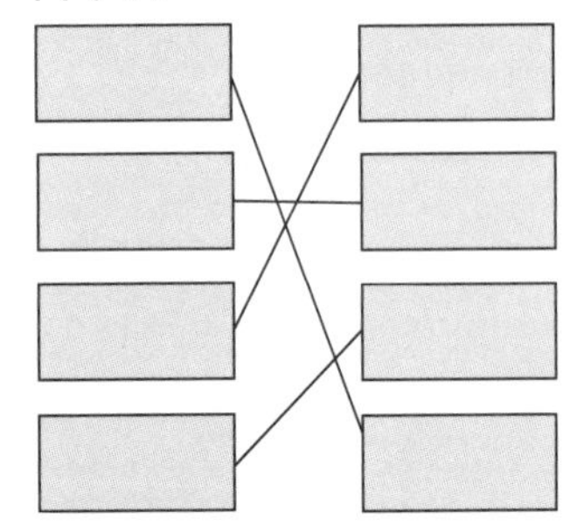
6. **(a)** For nuclei with a low proton number neutron number = proton number; As the nuclear mass increases extra neutrons are needed, so more massive nuclei have a greater neutron number than proton number
   (b) The alpha particle would not change the neutron proton ration (it has 2 protons and two neutrons) so would not make the low mass particles more stable.
7. **(a)** thermionic emission
   (b) the electrons need to gain energy to leave the metal; heating gives the electrons energy
   (c) KE = eV
   = 1.6 x 10-19 C x 4.2 kV = 6.72 x 10-16 J
8. **(a)** **(i)** +1.6 x 10-19 C ii) + 9.1 x 10 –31kg (+signs not required, but – is wrong)
   (b) They annihilate/destroy each other; leaving gamma rays
   (c) one which cannot be broken into smaller particles
   (d) down; -1/3e
   (e) KE = e V where e is the charge on the proton
   KE = 1.6 x 10 –19 C x 2500 kV = 4 x 10-13 J
   (f) saves money/allows them to share facilities and get more for their money/ share ideas/share expertise

Chapter 11

How Science Works

1. No it's the faster speed that causes more deaths. (There is a correlation but it is not the cause)
2. When the car is going faster it has more momentum and the time of the collision is shorter so the force on the pedestrian is greater, and this causes more injury.
3. The feel safer so will take more risk.
4. **(a)** drivers and passengers **(b)** pedestrians and cyclists
5. The argument that says the right decision is the one that leads to the best outcome for the majority of people involved.
6. Fines and speed cameras, and speed humps to reduce drivers speeds.

Answers to questions

1. C
2. B
3. B
4. **(a)** T **(b)** T **(c)** F **(d)** F **(e)** T **(f)** T
5. 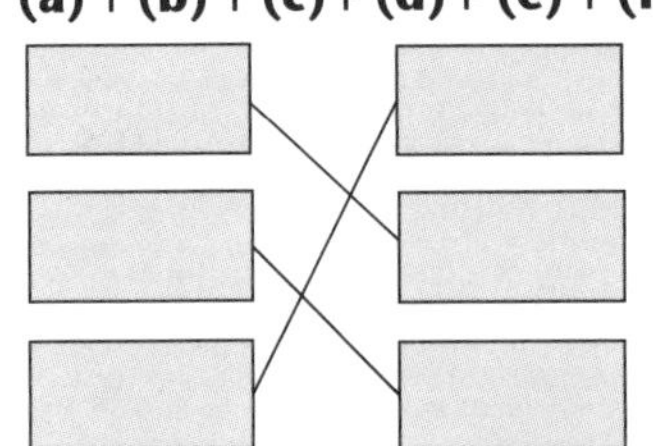
6. **(a)** **(i)** 20 x 1200 =24000 kg m/s
   (ii) 15 x 3000 = 45000 kg m/s
   (b) 24000 + 45000 = 69000 kg m/s
   (c) 69000 kg m/s
   (d) 69000 ÷ (1200 + 3000) = 16.4 m/s
7. **(a)** –3.8 x 10^{-25} kg m/s (must be negative)
   (b) 0
8. **(a)** gravity / gravitational force
   (b) towards Sun
   (c) 4F
   (d) faster because the force on it is larger
9. **(a)** 11N + 11N = 10N +5N + 2N + V
   V = 5N
   (b) 5N x x + 11N x 0.4m = 5N x 0.1m + 2N x 0.2 m+ 10N x 0.45m
   x = 1 ÷5 = 0.2m
   (c) **(i)** vase 3
   (ii) 1st reason: centre of mass is lower
   2nd reason: base is larger

Chapter 12

How Science Works

1. **(a)** T **(b)** F **(c)** T **(d)** F
2. Examples could be: Jodrell Bank, European Southern Observatory, Roque de los Muchachos in la Palma
3. Advantage: Avoids absorption and refraction effects of the atmsosphere
   Diadvantage: High cost of setting up (launch) maintaining and repairing/ Can be affected by problems with the space program (e.g. launch cancelled due to rocket design problems)
4. Astronomical: normal weather conditions/light pollution/altitude (higher better)
   Non-astronomical: cost/environmental impact/impact on local people/working conditions

Answers to questions

1. D
2. B
3. **(a)** F **(b)** T **(c)** T **(d)** F **(e)** T **(f)** T
4. A C D

5. Cepheid, variable, observed, brightness, period, intrinsic, brightness, period, intrinsic, brightness, intrinsic/observed, brightness, observed/intrinsic brightness (answers separated by / can be either first intrinsic then observed, or first observed then intrinsic)
6. **(a)** ✓ **(b)** ✓ **(c)** ✗ **(d)** ✓
7. **(a)** **(i)** away from bulb **(ii)** projector
   (b) changes the distance between the light sensitive detectors and the lens

Chapter 13

How Science Works

1. **(a)** Your answer should mention that some people (or people close to them) have serious/fatal medical conditions and they feel that a cure is more important than the lives of a few animals. Some people feel that so many people and animals will benefit that it is worth using the animals to find a cure.
   (b) Your answer should mention that some people think that animals have rights and should not be killed or harmed by humans.
   In both answers you should be objective and explain, 'What some people think' rather than stating your own view. Give your own view as well if the question asks for it.
2. Example answer:
   You have been asked to take part in a double-blind clinical trial of a new tracer for your scan. This means that neither you, nor your doctor will know whether you have been selected to use the new tracer or the one currently used for PET scans.
   The results of your scan and those of all the other patients in the trial will be studied to see whether the new tracer performs better than the old one, and if so it will be used for all patients. If it is not as good we will continue to use the old tracer. You can opt to take part in the trial or not. If you do not take part your scan will be done using the tracer currently in use.

Exam practice questions

1. D
2. D
3. B
4. 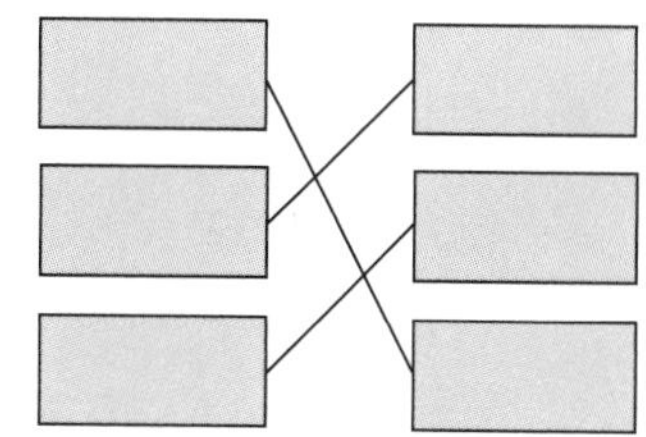
5. **(a)** F **(b)** T **(c)** T **(d)** F **(e)** F **(f)** T
6. 500Ω , 5V, 10Ω, 0.15V, dark, 1000Ω, resistor, variable, resistor
7. **(a)** step-down transformer
   (b) 12
   (c) diode
   (d) 4
   (e) see Fig. 13.15(c) page 208
   (f) stop it dropping to zero every half cycle by connecting a capacitor across the output terminals.
8. **(a)** They both use a gamma camera to record gamma rays/ In both cases patients are given a radioactive tracer.
   (b) The PET scan looks for pairs of gamma rays that arrive at the same time in opposite directions/ The tracer is a positron emitter not a gamma emitter.
   (c) By bombarding a stable isotope of an element with protons.
   (d) By tagging a molecule normally used by the body with a radioactive nucleus (for example replacing an oxygen atom in glucose with a radioactive fluorine nucleus) and injecting it into the bloodstream.
   (e) Looking for brain tumours/bone cancer/parts of the brain that are active when different activities are performed.
9. **(a)** **(i)** zero
   (ii) reversed
   (iii) zero
   (b) 1. faster rotation
   2. stronger magnetic field/magnet/iron core in coil to concentrate magnetic field
   3. more turns in the coil
   (c) slip-rings and brushes
   (d) $1 \div 0.025\ s = 40\ Hz$

Index